MANUFACTURING RESEARCH AND TECHNOLOGY 19

SOLID FREEFORM MANUFACTURING

MANUFACTURING RESEARCH AND TECHNOLOGY

Volume 1.	Flexible Manufacturing: Recent developments in FMS, Robotics CAD/CAM, CIM (edited by A. Raouf and S. I. Ahmad)
Volume 2.	Computer-Aided Design, Selection and Evaluation of Robots (B. O. Nnaji)
Volume 3.	Modelling and Design of Flexible Manufacturing Systems (edited by A. Kusiak)
Volume 4.	Flexible Manufacturing: Integrating technological and social innovation (P. T. Bolwijn, J. Boorsma, Q. H. van Breukelen, S. Brinkman and T. Kumpe)
Volume 5.	Proceedings of the Second ORSA/TIMS Conference on Flexible Manufacturing Systems: Operations research models and applications (edited by K. E. Stecke and R. Suri)
Volume 6.	Recent Developments in Production Research (edited by A. Mital)
Volume 7A.	Intelligent Manufacturing Systems I (edited by V. R. Milačić)
Volume 7B.	Intelligent Manufacturing Systems II (edited by V. R. Milačić)
Volume 8.	Proceedings of the Third ORSA/TIMS Conference on Flexible Manufacturing Systems: Operations research models and applications (edited by K. E. Stecke and R. Suri)
Volume 9.	Justification Methods for Computer Integrated Manufacturing Systems: Planning, design justification, and costing (edited by H. R. Parsaei, T. L. Ward and W. Karwowski)
Volume 10.	Manufacturing Planning and Control – A Reference Model (F. P. M. Biemans)
Volume 11.	Production Control – A Structural and Design Oriented Approach (J.W. M. Bertrand, J. C. Wortmann and J. Wijngaard)
Volume 12.	Just-in-Time Manufacturing Systems – Operational planning and control issues (edited by A. Şatır)
Volume 13.	Modelling Product Structures by Generic Bills-of-Materials (E. A. van Veen)
Volume 14.	Economic and Financial Justification of Advanced Manufacturing Technologies (edited by H.R. Parsaei, T.R. Hanley and W.G. Sullivan)
Volume 15.	Integrated Discrete Production Control: Analysis and Synthesis – A View based on GRAI-Nets (L. Pun)
Volume 16.	Advances in Factories of the Future, CIM and Robotics (edited by M. Cotsaftis and F. Vernadat)
Volume 17.	Global Manufacturing Practices – A Worldwide Survey of Practices in Production Planning and Control (edited by D.C. Whybark and G. Vastag)
Volume 18.	Modern Tools for Manufacturing Systems (edited by R. Zurawski and T. S. Dillon)
Volume 19.	Solid Freeform Manufacturing – Advanced Rapid Prototyping (D. Kochan)

MANUFACTURING RESEARCH AND TECHNOLOGY 19

Solid Freeform Manufacturing

Advanced Rapid Prototyping

Edited by

D. Kochan

Technische Universität Dresden, Institut für Produktionstechnik

Dresden, Germany

ELSEVIER

Amsterdam – London – New York – Tokyo 1993

ELSEVIER SCIENCE PUBLISHERS B.V.
Sara Burgerhartstraat 25
P.O. Box 211, 1000 AE Amsterdam, The Netherlands

Library of Congress Cataloging-in-Publication Data

Kochan, Detlef.
Solid freeform manufacturing : advanced rapid prototyping / D. Kochan.
p. cm. -- (Manufacturing research and technology ; 19)
Includes bibliographical references and index.
ISBN 0-444-89652-X (acid-free paper)
1. CAD/CAM systems. 2. Solid freeform fabrication. 3. Prototypes, Engineering. I. Title. II. Series.
TS155.6.K683 1993
670'.285--dc20 93-31067
CIP

ISBN: 0 444 89652 X

This book is printed on acid-free paper.

Transferred to digital printing 2005

Foreword

All industrial development is driven by new inventions. The time span between the first functional products and broad industrial application differs, and depends on many factors. In this case, the first research results concerning solid freeform manufacturing were published in 1982, and industrial application started in 1989 with the Stereolithography system from the 3-D Systems Corporation in California, USA.

At the end of 1992, 12 different kinds of equipment will be commercially available and more than 20 procedures patented or under development. These facts demonstrate the dynamic development in this new "high-tech" area. In many articles and conference proceedings, the new procedures are explained from different perspectives.

This book Solid Freeform Manufacturing should contribute to an acceleration of industrial applications by providing a good explanation of the new technological methodologies including:

- characteristics of the most suitable areas of application in manufacturing industry
- systematic approaches to decision-making for the most suitable equipment and usages (whether to invest or farm out)
- methodical support for proper utilization of selected equipment
- efficient combination with current manufacturing technologies
- technological and economic aspects for evalution and comparison of other traditional procedures
- systematic investigation of new application areas

The book is primarily addressed to interested users in all important industrial fields such as:

- CAD / CAM
- Marketing

- Engineering
- Manufacturing (especially tool making)

In line with the typical application areas, the book is also suitable for students of engineering, management, and economic disciplines.

Acknowledgements

All my in depth studies concerning Solid Freeform Manufacturing were supported worldwide by scientists and experts from universities, research centres, and vendor and customers company involved with this new technological discipline.

I would like to thank Prof. Björke and his colleagues, especially R. Hovtun and N. Ruud, for providing worthwile experience and support during my sabbatical period (January to June 1992) at NTH SINTEF, Trondheim, Norway.

Practical tests were supported by BIBA Bremen (Prof. Hirsch and co-worker) and 3-D Systems at Darmstadt, Germany.

Much informative material was obtained from and interesting discussions held with, experienced experts of vendor companies including

- C-MET / Japan
- Cubital / Israel-Germany
- Dornier / Germany
- D-MEC / Japan
- DTM / USA
- DuPont / USA
- EOS / Germany
- Helysis / USA
- Light-Sculpting / USA
- Stratasys / USA
- 3-D Systems / Germany-USA

In the framework of joint activity with the Numerical Control Society of Germany, a specific challenge test-part was created; the practical realization was supported by leading technological companies.

Well known experts of predominant technological companies such as:

- BMW / Germany
- Chrysler Corporation / USA
- Dornier / Germany
- Ford Motor / USA
- General Motors / USA
- and others

organized informational visits and delivered informative application experiences.

During my international scholarship at IAMS-Cincinnati (Autumn 1992) I enjoyed the creative working atmosphere and worthwhile discussions, especially with Dr. Gene Merchant, Mr. John Kohls, Ms. Susan Moehring, Mr. Steve Curfiss and others.

Interesting discussions with experts of advancded universities led me to decide to prepare a suitable book in this new technological area.

For the preparation of my book, including text and graphics, I would like to thank Mrs. L. Hermann, Mrs. B. Proft, TU Dresden, Germany; Liv Sodal, NTH Trondheim, Norway; and Karen Barber and Lori Ortega, IAMS, Cincinnati, Ohio, USA.

Last, but not least, I would like to thank most, cordially the Walter and Gisela Gildemeister family for providing me with a second home while in Cincinnati.

Contents

1. Introduction

Computer-aided generation of geometrical objects by the application of specific physical principles with new chemical materials signified a drastic change in advanced manufacturing technology. The traditional vision of the "Factory of the Future" by increased automation of all necessary functions like machining, handling, assembling,needs to be reconsidered.

Several hundred examples of practical applied equipment, which are able to manufacture complicated geometrical objects without conventional tools in one or a few operations, are the evidence of a changed vision of the "Factory of the Future."

The invention of NC (Numerically Controlled) technologies in the midnineteen- fifties (1953) marked the beginning of a new kind of flexible manufacturing. In the development of advanced manufacturing technologies it was characterized as a "watershed event" /1/.

This first step of practical computer application led to a broad change of automated manufacturing technology.

The NC technology is nowadays used for different manufacturing procedures, realized in various single machine tools, machining centres and manufacturing systems.

In parallel with the developments of NC programming languages, more and more engineering tasks were also supported by the application of computers. One essential explanation for the general success of NC technologies is the possible connections between numerical output devices and numerical planning processes.

The current state of development in geometrical modelling can be characterized by the increased application of 3 D CAD systems. This is the basic and prerequisite for the new developments in rapid prototyping, solid freeform manufacturing, layer manufacturing technologies, and others similarly named. The practical step in this direction was made possible with the invention of specific polymer liquids, which are hardened by laser beam or UV light. These facts and possibilities arouse realistic expectation of the new area of manufacturing technologies. Its expected success and broad application may be explained as follow:

- It matches a numerical output device to a numeric design process.
- It makes possible a flexible automation of forming, casting and model-making processes for multi-sided applications /2/.

Apart from the normal start-up-difficulties of such general new technologies, the current developmental stage can be recognized as the beginning of a new area of manufacturing technologies.

The best application results can be achieved if the new technologies become integrated into an advanced "high--tech" environment /3/4/.

This book therefore starts by considering the new era relationships to advanced flexible automation, especially NC technologies (Chapter 2). For the sake of general application orientation, the essential commercially available solid freeform manufacturing systems are explained in a general overview (Chapter 3).

It should be pointed out that new application areas will be investigated in the near future. These are all scientific, industrial or medical working spheres in which 3-D data will be modelled. The medical application field is under development, and other disciplines will soon follow.

In this book, however, the application of manufacturing technology will be emphasized.

The description of the most readily available principle (Chapters 4,5,6), with the essential features like the accuracy, material behaviour, costs, advantages and disadvantages, will allow different interested experts to develop their own approaches.

The current supremacy of Stereolithography principles, which lead on to most of the available experiences in this field, is the reason for their detailed description. But it should be emphasized that a lot of synonymous solutions will become available. Therefore, some support is given for the selections of the most suitable equipment for chosen requirements, which is also a considerable task.

The overview of current experiences in essential application areas will stimulate interested users (Chapter 8). Related to the specific properties of the usable materials, further explanations are given in Chapter 9.

For better evaluation of the most suitable manufacturing processes for freeform machining, the most closely related manufacturing processes (multi-axis milling) are characterized in Chapter 10.

Concerning the commercially available procedures, the main points for decision-making and essential economic aspects will be explained in Chapter 11.

In Chapter 12, some important development trends are briefly outlined. In view of the entire dynamic development in the new discipline, continuous study of the literature and conference proceedings is essential. But the book will provide a foundation for all necessary activities.

2. Advanced Manufacturing Technology: The State of the Art

An essential development step towards a higher degree of automation was the invention of NC (numerical control) technology in the mid- 1950s. But the spread into a broad distribution within industry, beginning in the mid-1960s (1965), was relatively slow. The usually start- up difficulties, and the relatively high costs of machine tools, controllers, programming and education of the users, were reasons for this slow application process during the first 10 years. It is particulary the drastic reduction of prices following the introduction of microelectronics, the user support offered by NC programming languages, increased reliability and some other factors which have led to a broad application of NC technology in the last 20 years.

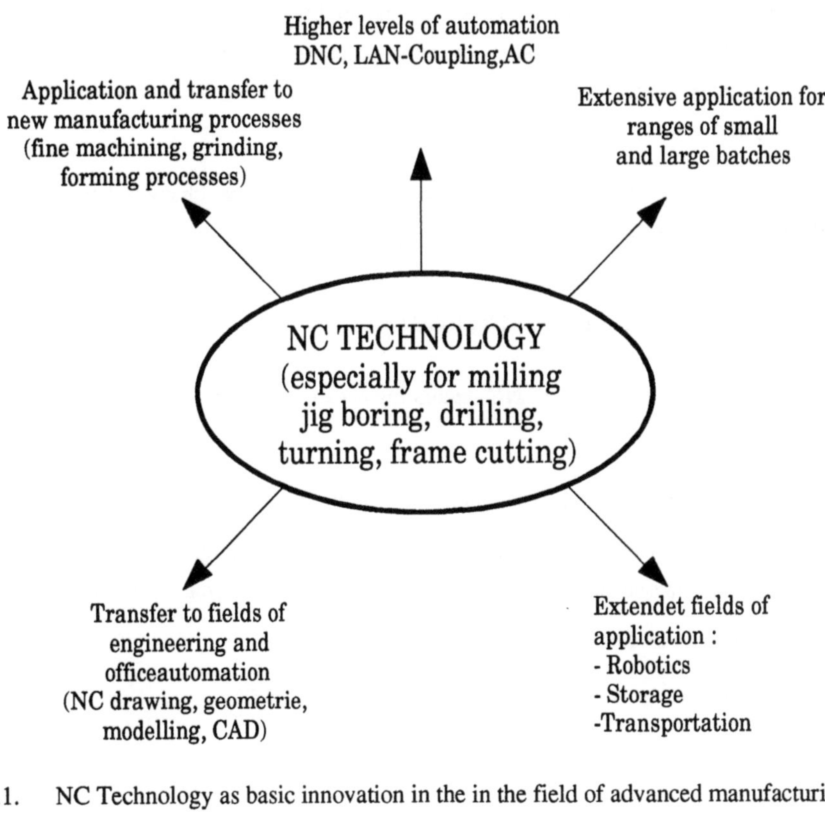

Fig. 2.1. NC Technology as basic innovation in the in the field of advanced manufacturing

Based on its first main fields, of milling, turning and boring, the NC technique is now widely used in many different areas (see Fig. 2.1). The influence of NC technology can be characterized by developments

- towards higher degrees of automation
- in the direction of flexible automation in mass production
- in the direction of flexible automation for one-of-a-kind production
- in different application fields such as
    - grinding finishing-machining with integrated quality control
    - sheet-metal manufacturing
    - robotics
    - material handling, transport and storage

Essential features of NC technology like speed and highly-accurate feed control by powerful controllers are the basis for inclusion of new procedures, for example laser technology (laser-cutting of sheet metal, hardening of workpieces, laser-caving of new materials which are difficult to cut) and others.

For further explanation of the current development stage, some characteristics of complex machining will now be describet.

2.1 Complex machining with NC technology /4/

The current development stage in NC technology can be characterized by the increased application of

- NC machine tools for different procedures and operations
- machining centres for complex machining by milling, boring and other operations - clamping of prismatic parts in one clamping position and machining in different directions
- machining cells for all kinds of cutting processes
- flexible manufacturing systems for complex process automation.

The state of the art in NC technology is featured by machining cells with varied automated functions and possibilities for process supervising, including adaptive control (see Fig. 2.2).

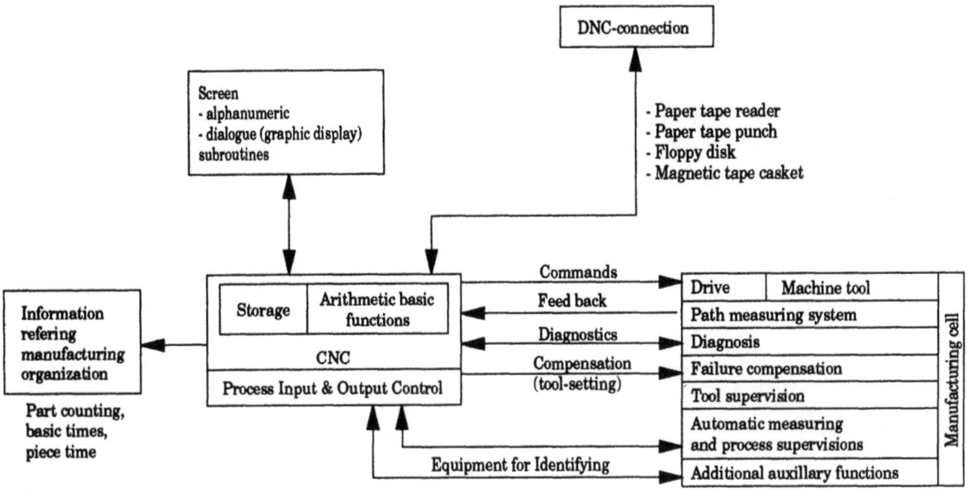

Fig. 2.2. Principle of including Adaptive Control (AC)

New approaches to machining cells for multiple-operations have been developed in past years, especially in connection with turning cells. The basic functional have been extended by:

- multi-spindle processing with different tools,
- multi-tool processing for turning operations,
- processing with controlled c and y axes, which allows additional boring and milling operations in one clamping position
- rear drilling attachment.

The implementation of such complex processes, controlled in 4, 5, 7 or more axes, required specific new approaches for process planning and intelligent software support.

The developmental stage of process automation, in various kinds of Flexible Manufacturing Systems (FMS), requires the automation of many functions of the entire production process such as:

- machining in different operations,
- tool changing, management and monitoring,

- workpiece handling, storage and transport,
- supervising, quality control and other auxiliary functions.

It should be pointed out that from the technical point of view, flexible manufacturing systems are the core of computer-integrated manufacturing (CIM).

It is not intended to explain CIM strategies and requirements in detail, but for better understanding of the tremendous expenditure in Hardware, Software and Manpower for installation and efficient utilization of CIM-relevant systems, Fig. 2.3, will give some impressions.

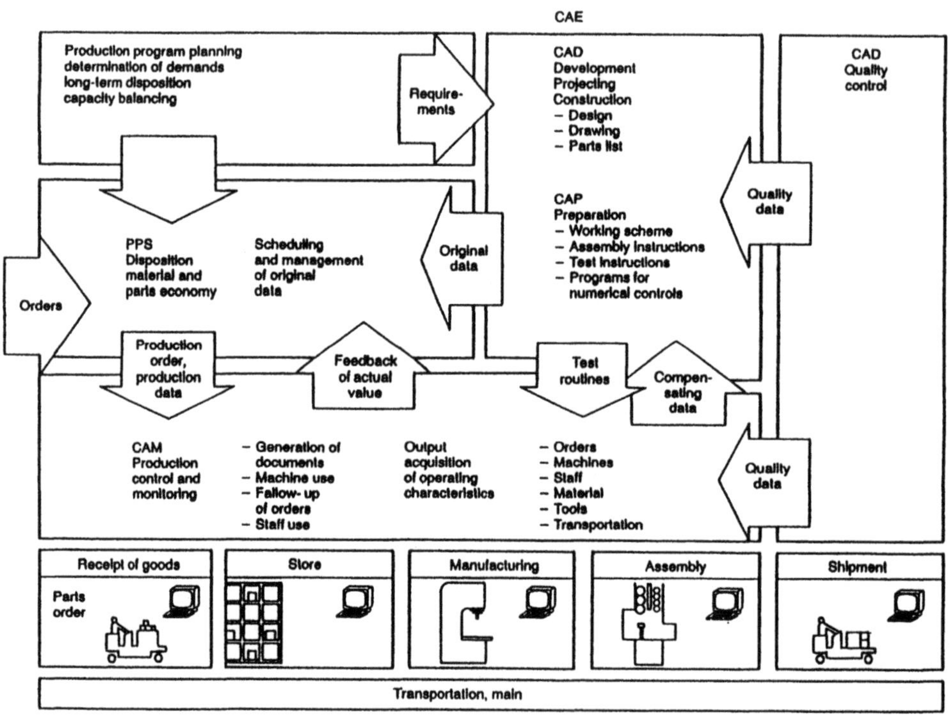

Fig. 2.3. Main functions of Computer-integrated Manufacturing (from Waller - SIEMENS)

From this background to the state of the art in advanced NC technology, the general advantages of the new Solid Freeform Manufacturing Technologies will be considered in Section 2.4.

For an explanation of the principal similarities and differences, Section 2.3 gives some general information.

2.2 Application of new physical principles

The whole history of technical development is influenced by synergy effects, meaning that important progress to new technological levels comes about mainly by integration of specific effects of different branches.

The basis for the advent of Solid Freeform Manufacturing is a combination of different disciplines, in this case

- Advanced computer technology
- NC technology
- Laser technology, and other physics principles
- newly available materials (see Fig. 2.4)

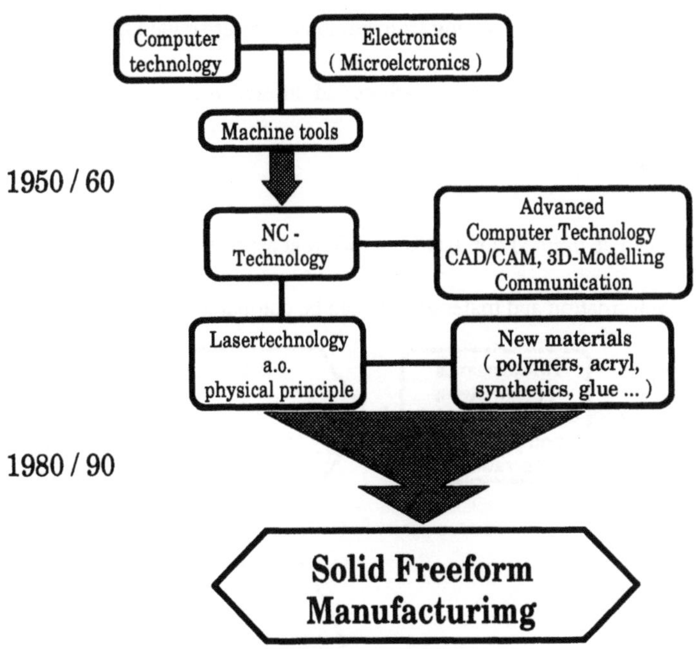

Fig. 2.4. Development steps in advanced manufacturing technology

A prerequisite for all further steps is the availability of 3-D models in connection with extensive data transfers via advanced communication networks. The NC technique is necessary for high-precision control of quick movements.

Based on the available new materials, some specific physics principles can be applied, like

- polymerization of suitable resins by laser or other light beams
- selective solidification of solid particles (powder) by laser beams
- binding of liquid or solid particles by glueing or welding
- melting and resolidification.

2.3 Comparison of NC Technology and Solid Freeform Manufacturing

It is not easy to compare technical solutions at different stages of development, but because two tendencises can be observed

- the continuous further developments of well known principles
- the realization of new effects arising from combinations of single known principles

it seems interesting to distinguish the similarities and differences.

2.3.1 Similarities

Some similarities can be pointed out, based on different aspects:

- The general information and material flow can be compared (see Fig. 2.5 and 2.6)

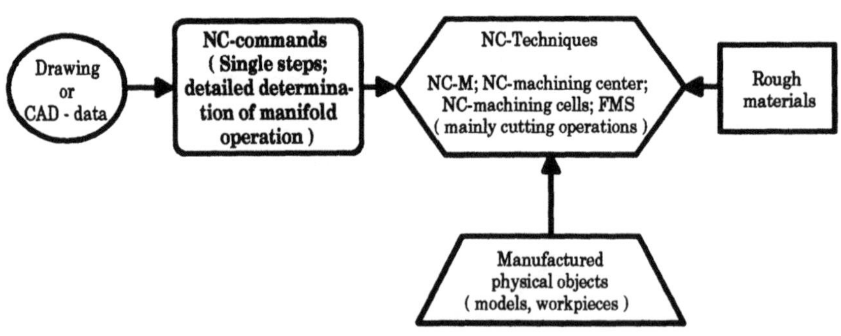

Fig. 2.5. Principle NC-Technology

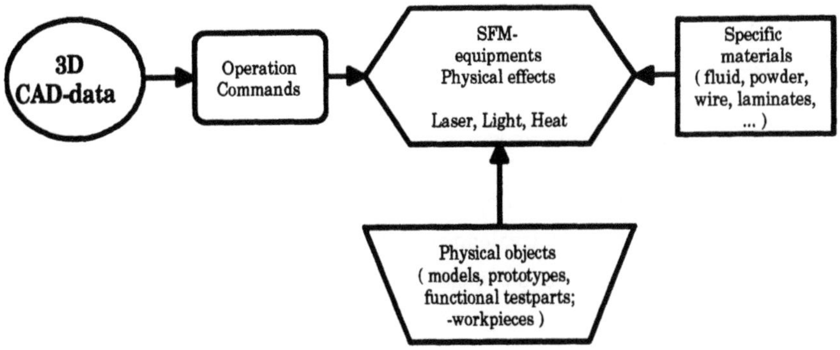

Fig. 2.6. Principle of Solid Freeform Manufacturing

- The ability of the equipment to understand numerical information as a control input
- This ability includes high flexibility as a main feature
- The practical mastering of all possible procedures requires highly qualified people.

2.3.2 Differences

Based on the characteristic similarities, the following features mark the essential differences:

2.3.2.1 Data and input preparation

For NC technology, a drawing or 2-D CAD data can be used more or less as they are delivered by the designer. Only a few additions of manufacturing information are necessary (definition of coordinate systems, clamping positions and devices, definition of tool path, etc.)

For the SFM technology, very accurate 3-D models are necessary. A description in wireframe models is not sufficient. What is required are exact solid or surface models, with absolute closed volumes.

The preparation of NC commands for detailed description of all necessary operations, tool movements, tools, working values, and so on required a tremendous expenditure, which is partly ameliorated by suitable NC programming systems.

The preparation of operating commands for SFM is reduced mainly by geometrical information, well supported by graphics workstations, but it has to be carried out very carefully and with the benefit of specific experiences.

Summarizing these characteristic differences, the following point can be made.

In the application of NC technologies, important tasks were transmitted from the skilled worker to the NC programmer. In the next development step, a further transmission of essential tasks to the design area is required. Based on initially designed 3-D data models, the manufacturing engineer is responsible for the final data preparation in accordance with the requirements of the SFM principle.

The creation of 3-D models starting from drawings is a time- consuming task. The best possible effects can therefore be achieved if the 3-D models are prepared and can be used for further tasks (e.g. FEM analysis, flow analysis for injection moulding, NC programming, etc.)

2.3.2.2 Process controlling and realization

All automated procedures at NC equipment are associated with high technical expenditure. Unattended manufacturing is only realized in some cases. Flexible automation applies mainly to cutting processes. For the entire process, involving many technical aids (tools, fixtures, chucks, tool-changer and magazines, handling and transportation equipment, etc.) some sophisticated control software is necessary.

The SFM procedures are highly automated, and unattended operation is normal. Additional tools, chucks, and specific equipment for material handling and transportation between machining stations and so on are not necessary. Complicated operations are automated in one-step procedure.

2.3.2.3 Material aspects

The different currently usable materials are one of the most important differences (see Chapter 9). For the NC technology, all industrial technical materials, mainly metals, are in use. This allows the manufacture of all technical workpieces and products without restrictions.

The SFM technology is currently restricted to specific materials. Only in a few cases can workpieces from plastic materials be produced in small series runs. The physical objects so made are used mainly as prototyps, models, functional test parts or patterns for other procedures.

Independent of these characteristic restrictions, some advantages can be enumerated.

2.4 Importance of Solid Freeform Manufacturing

The hig importance arises from possible improvement in all application areas (see Chapter 8). In addition, some general aspects must be considered.

2.4.1 CIM-adapted procedure

For a general evaluation of the importance of Solid Freeform Manufacturing technology, it seems useful to start with the comparable features of advanced NC technology.

The NC technology was applied to traditional machine tools, which were further development for this reason.

The SFM technologies have been fully development in tandem with CAD/CAM and CIM requirements.

These facts lead to drastic cost reductions in comparison with currently known CIM requirements.

The most important aspects are described by Kruth /5/ as follows.

Automating the whole process from design to manufacture causes tremendous problems, such as knowledge capture and computer handling of knowledge (expert systems), planning (CAPP), interfaces (STEP, MAP, ...) and communication.

The fact that Solid Freeform Manufacturing only creates material where needed and without any tooling, makes it well suited for CIM and eliminates most of the problems encountered with other processes:

- no need for feature-based design: a 3-D surface or solid model of the product is sufficient, because the manufacturing process does not have to be adapted to the type of geometry or feature;

- no need for conversion from design to manufacturing features: the design contains all the geometrical information needed for manufacturing, because there is no difference between the part's material and the material to be machined;
- no need to define a blank geometry;
- process and operation planning are reduced to a minimum;
- no need to define sequences of operation (CAPP) and complex handling, as the part is produced in one operation;
- no problem of defining different set-ups: the part is made in one set-up;
- no need for clamping, jigs or fixtures (however, some processes may require that a support structure is built up together with the workpieces);
- tool-less processes: no need for design and manufacture of moulds and dies.
- etc."

Solid Freeform Manufacturing allows a direct interface from CAD to CAM. It almost completely eliminates the need for CAPP.

This leads to tremendous reduction of all necessary expenditure for the manufacturing of physical objects (manpower, technical equipment and software aids).

2.4.2 SFM in an Integrated Product Development environment (Concurrent Engineering)

The integrated product development methodology allows designers and manufacturing engineers to communicate early in the design phase of a product. The key concepts and advantages of Integrated Product Development are as follows /6/

1. Attain a balance between product and process technology (understand the planning, manufacturing and support of a new design while creating the design).

2. Use multi-discipline teams that are empowered, to make critical decisions promptly whenever necessary for the benefit of the product and its potential cost.

3. Foster more effective communication and information transfer among the many different people who are involved in the design, production and support of the final products (e.g. aerospace, automotive or other products).

Solid Freeform Manufacturing is one way to help decrease a product's development time and reduce its life cycle cost.

2.4.3 Relations with the traditional manufacturing processes

It seems clear that the new Solid Freeform Manufacturing technique cannot replace traditional manufacturing processes.

In addition to both characteristic effects (relevance to CIM and to integrated product modelling), it is most important to note that in some combinations with other traditional processes remarkable advantages can be attained.

The general aspect of prototyping without any other manufacturing procedure was explained previously.

A significant advantage is provided by the combination of pattern- making and casting procedures. It should be emphasized that SFM techniques allow a flexible manufacturing in casting processes. These include sand casting, investment casting, soft tooling, and similar procedures.

Up to now no combination with forming procedures is known.

In connection with cutting operations, it is possible to use models for copying. Much more important for many products is the assembly testing. It can be expected that such application will also be a very important application field for SFM techniques. Another most important application area is related to coating procedures, for example the well known HCP metal-spraying technology using MCP low melting alloys (~220°C melting point) successfully applied in connection with SFM objects. Initial experiments on the coating of copper materials with the aim of manufacturing cathodes for EDM procedures are under development. But in general, the combination with different traditional manufacturing procedures is an essential working field, and will be explained with typical examples in Chapter 9.

3. Solid Freeform Manufacturing

3.0 Historical development

Based on the availability of 3-D geometrical modelling system, attempts at the direct generation of physical objects, without traditional tools, were begun in the 1970s. In the 1980s the physical prerequisite became available with the specific hardening feature provided by a laser beam. The first independent experiments in USA, Japan and Europe led to technically realized solutions /7/8/9/. Because of the international publication of the idea from Kodama (Japan), no international process patents are possible in this field. In 1988, at the US Autofact Fair, the US enterprise 3-D Systems exhibited the first industrially usable technical equipment under the description "Stereolithography Apparatus (SLA)". At the end of 1992, Stereolithography equipment was in industrial use in more than 500 cases, in 17 countries. There are known to be 12 other examples of commercial equipment available (8 using the similar Stereolithography principle), and more than 30

COMPANY NAME	PRINCIPLE	PRODUCT
3D Systems, Inc. USA	Stereolithography	SLA-190, SLA-250, SLA-500
DuPont USA License to Teijin Seiki Japan	Stereolithography	SOLJFORM
CMET Japan	Stereolithography	SOUP 400 GA/GH, 530 RH, 600 PA/PH, 850 PA
D-MEC Japan	Stereolithography	Sony JSC-1000, JSC-2000, JSC-3000 Solid Creation System
Mitsub. Eng. Japan	Stereolithography	COLAMN
EOS Germany	Stereolithography	STEREOS 300, 400, 600

patented solutions (see Fig. 3.1). But independent of this tremendous progress, it has to be pointed out that there are many problems to be solved in coming years before this new technology is in common use.

COMPANY NAME	PRINCIPLE	PRODUCT
Cubital Israel	Lighting of Plans	Solider 500
Light Sculping USA	Lighting of Plans	LSF-0609 MA Photosolidification System
DTM-Corp. USA	Selective Laser Sintering	Sinterstation 2000
Stratasys, Inc. USA	Fused Deposition Modeling	3D Modeler
Helysis, Inc. USA	Laminated Object Manufacturing	LOM-1015, LOM-2030
Sparx, AB Sweden	Plastic Hot Plot	Hot Plot

Fig. 3.1. Equipment vendors and commercial available Systems 1993

3.1 Principle and terms

The very general principle is shown in Fig. 3.2, which demonstrates the following points.

1. The essential prerequisite of 3-D Data models
2. The application of specific materials such as fluids, powder, wire or laminates. This is currently different from traditional manufacturing processes, a decisive restriction
3. The use of newly developed highly sophisticated equipment employing different physical principles
4. The physical objects are currently, limited in size depending on the working ranges at the SFM equipment, but without restrictions concerning complexity and geometrical features.

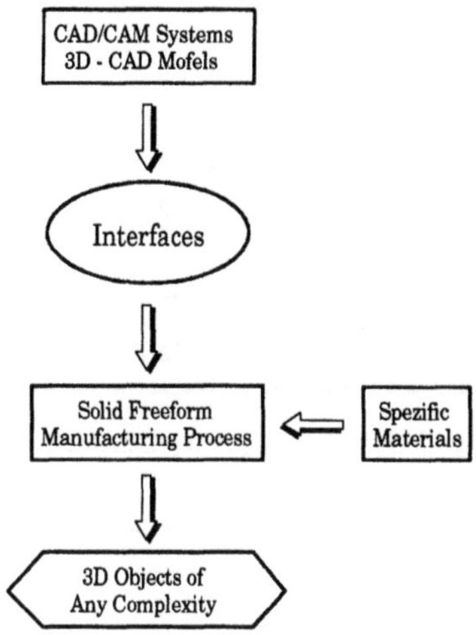

Fig. 3.2. Simplified Principle of Solid Freeform Manufacturing

The traditional subdivision of manufacturing technology is given by the following procedures:

- shaping by casting (powder metallurgy processes)
- forming
- cutting
- joining (welding, etc.) and assembling
- coating
- changing of substance properties

It should be pointed out that at least three of the main features mentioned

- forming
- coating
- changing of substance properties

are suitable for further examination in the new procedures. So on the basis of different technical disciplines like

- Computer technology
- Physical principles
- Chemical materials

it is worth looking at the details. In connection with the development of different

it is worth looking at the details. In connection with the development of different technical solutions, some definitions are in common use (see Table 3.1). A detailed comparison of terms used was made by Kruth /5/. Table 3.1 includes only some essential terms. Based on the many and different terms currently used, it seems necessary to look for one suitable short definition. The comments in Table 3.1 are the starting points for the following proposal.

Term	Comments
CAD oriented Manufacturing	Not only CAD oriented (every manufacturing technology has to be customer oriented!)
Direct CAD Manufacturing	See above
Desktop Manufacturing	Not general to accept; the desk could be only a very small facility for manufacturing
Instant Manufacturing	In every manufacturing process will be produced "instand" parts
Layer Manufacturing	Characterized most, but not all of the new principles
Material Deposit Manufacturing	Includes the geometrical, not the physical aspects
Material Addition Manufacturing	see above (MDM)
Material Incress Manufacturing	Good explanation also from the geometrical aspect; difficult to translate
Solid Freeform Fabrication Solid Freeform Manufacturing	Both terms includes the the aspect of solidification and the geometrical complexity of the manufacturing procedure
3D Printing	Printing is only one possibility
Rapid Prototyping	Widely used, but also in computer technology with other content

Table 3.1. Essential Terms for Solid Freeform Manufacturing

The first it is to emphasize that a manufacturing procedure will be characterized, that is in agreement with most of the currently used terms. Furthermore, it seems important to include the aspect of solidification of different materials and the possibility of creatin physical objects of any complexity.

Therefore the following term will be proposed: **Solid Freeform Manufacturing.**

The term, Solid Freeform Manufacturing seems to be the most suitable because :

- solidification is a typical feature of all the different procedures
- freeform is the characteristic term for geometrical objects of any complexity
- all kind of the new procedures are unique manufacturing processes.

3.2 Features and possibilities for classification

For all currently known SFM procedures, the unified starting point is a computer-held 3-D volume or surface model, cut open in small slices. The x-y surface elements created are the basis for controlling appropriate equipment (laser beam, optical glass plates, heating nozzles, ink-jet printer principle, etc.). Based on general similarities of the 3-D modelling of any kind of objects for the characteristics of the different procedures, the following features can be pointed out:

- application of different lights (illumination point by point, or surfaces)
- use of different energy sources
- differentiation by typical application ranges

The essential possibilities will be discussed below.

3.2.1 Division by lighting of photopolymers

The hardening by illumination of photopolymers is currently the most used procedure. The necessary materials are available from different companies in USA, Japan and Europe (see Chapters 4, 8 and 9). Based on this given situation, two principles are possible and are realized:

- lighting point by point
- lighting of plan surfaces

Some essential principles are shown in Fig. 3.3, which demonstrates the dominance of the laser application in Stereolithography-like procedures. Therefore this principle will be the focus of further explanations.

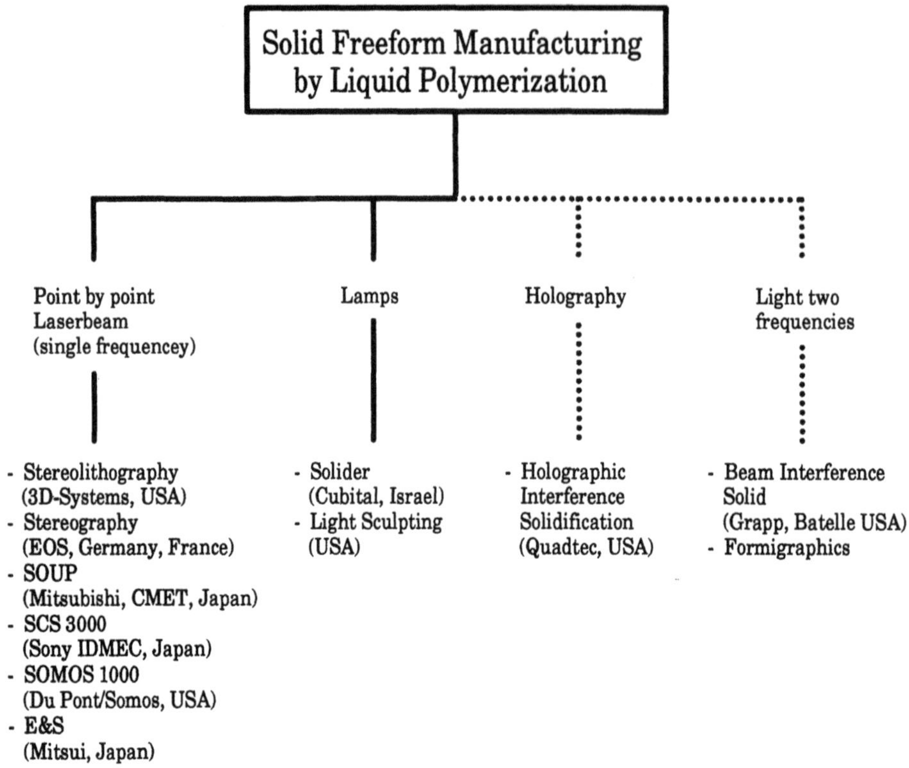

Fig. 3.3. Commercialized (_____) and Patented (_ _ _ _) Systems for Liquid Polymerization

3.2.2 Division by materials used

The division by materials used is important because the current mainly used liquids do not have the typical features of traditional mechanical materials. The subdivision in fluids, powder and solid materials (mostly by kind of laminates, foils or wires) is demonstrated in Fig. 3.4.

Based on powder applications, two principles are currently developed:

- Selective Laser Sintering (SLS), which is realized by hardening of layers of powder by a controlled laser beam
- 3-D printing, which uses the ink-jet-printing principle with glue. The glue is sprayed in a small beam onto thin layers of powder, which will be bound and hardened.

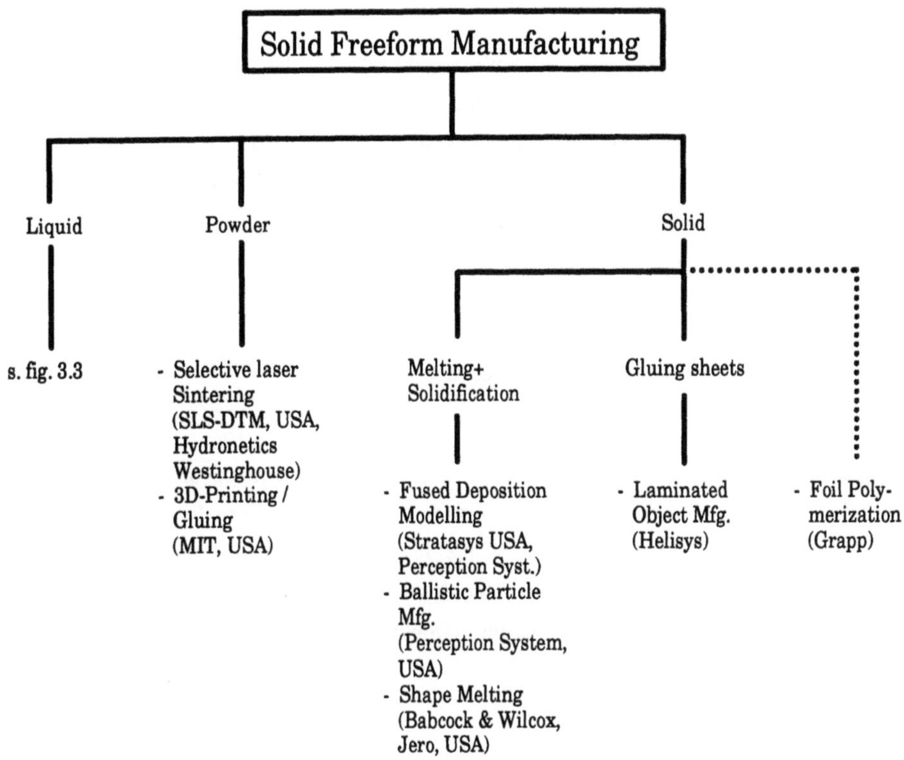

Fig. 3.4. Division of SFM-techniques by materials use

The application of solid materials is realized with plastic or nylon, as wire or foils:

- Wire-like materials are softened by heating nozzles and continuously built up to a complex object, layer by layer
- Laminates are glued, layer by layer, and then the superfluous outside or inside contours are cut by a controlled laser beam.

3.2.3 Division by energy used

This possibility of classification is more or less included in the above-mentioned principles (see Fig. 3.4). Therefore no detailed assessment of the different available solutions is included in the overview. It should be pointed out that the best possible accuracy with a heat nozzle is much less than with a laser beam or lamps. The application ranges are therefore restricted to the evaluation of designs (see Section 3.2.4). One exception is the shape melting of large parts, but this is not a real SFM process.

3.2.4 Division by typical application ranges

This essential subdivision is only indirectly connected with the different procedures and principles, and mainly concerns its results, the generated physical objects. The possible application ranges are determined by some physical properties such as elongation, hardness and tensile strength, and especially by the accuracy. Based on these requirements, subdivision into four main areas is possible

- prototypes for evaluation of design
- prototypes for evaluation of functions
- models for further manufacturing processes
- functional parts for small series

It should be pointed out that all these application areas are of practical importance. The first applications were directed to the evaluation of designed objects. In some cases, only this support for the designer justified the investment. But the broad extension of application ranges into manufacturing is the driving force for most further developments. In general for all kinds of applications, the following essential advantages can be pointed out:

- significant time savings
- improvement to superior design
- communication-enhancing bridge between design and manufacturing.

3.2.4.1 Prototypes for evaluations of design

The current state of the art of geometrical modelling gives the designer great support in such functions as:

- viewing
- shadowing
- tilting
- magnifying, and others.

But most important for a realistic evaluation of a designed object is the further possibility of touching and feeling.In particular, aesthetic design requires a physical object in every case. The classical approach for building objects is by external shapes. These data are not usable for further computerized processing without time-consuming digitizing procedures. Any kind of interior shapes of designed parts (for example, consumer goods like cases) cannot be modelled in such a manual way. Therefore the same manufacturing procedures as used for batch- or series-production are necessary. This is time and cost-consuming costly, and normally requires weeks or months. From the point of view of competitiveness, rapid prototyping is an essential factor. Especially for marketing requirements, it is quite useful to act with real objects (marketing samples).

3.2.4.2 Prototypes for functional tests and evaluation

In addition to the evaluation of designed parts, it is necessary to generate test objects for specific functions or assembly. Such functional test can be fulfilled, for example a model of a car body in a windtunnel. Turbine blades or propellers can be tested for their degree of effectivness. For all kinds of consumer goods, it is necessary to check if the housing or cases fit together with other inside parts. That means that for all requirements of assembly tests, the application of suitable models is useful.

3.2.4.3 Models for further manufacturing processes

Apart from the breviously explained applications for

- visual representations for design evaluation and marketing samples
- functional durable prototypes

it should be noted that the generated models can be used as patterns for different manufacturing processes, especially for:

- investment casting
- vacuum casting
- metal spraying
- EDM

Some principles, like SLS and Cubital, are able to create wax patterns for investment casting. This is one main way to connect with the traditional metal manufacturing. On the other hand, all known SFM procedures are able to create plastic patterns. By coupling and combining with such advanced manufacturing technologies as soft tooling and spray metal tooling, it is possible to produce hundreds or thousands of parts starting with a single model .

4. Industrial Application of SFM Techniques

Of the different modes of subdivision of the entire working area, as described in Section 3.2.1, the division by materials used seems the most important. All other features are related to the material aspects, so further explanations will consider this major point. Most of the current industrially applied systems use photopolymer materials, to be hardened by laser beams. Therefore the further detailed explanations will start with this typical procedure.

4.1 Point-by-point layer solidification: Stereolithography

4.1.1 General principle

The principle of Stereolithography (point-by-point layer solidification) is outlined in Fig. 4.1. Based on this general sequence, the entire series of working steps can be subdivided into three main areas:

- Preparation tasks, to be carried out at the CAD workstation or SFM computer
- SFM procedure for layer solidification
- final works, such as cleaning and resin removal (postprocessing)

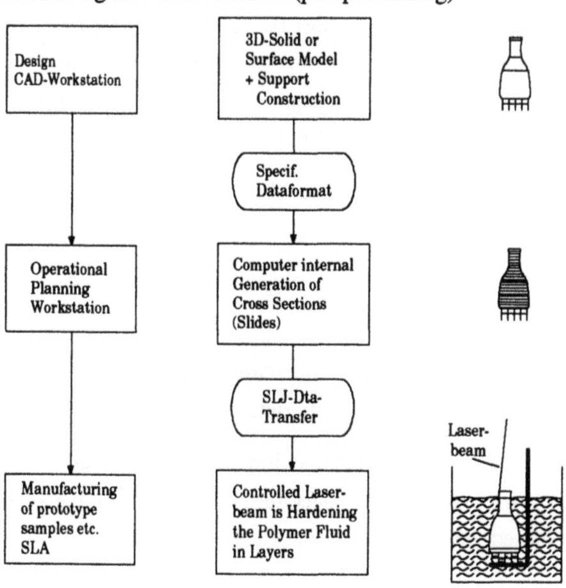

Fig. 4.1. General principle of Stereolithography

4.1.2 Preparation tasks

Geometrical modelling (see also Chapter 7)

The initial stage for point-by-point solidification is a 3-D surface or solid model. This data must represent a clearly defined enclosed volume /10/. For geometrical modelling there are nowadays different CAD/CAM-systems available.

A prerequisite for further steps is the necessary standardized Interface, as currently provided by most CAD suppliers. The CAD modellist needs a great deal of practice and experience in Stereolithography processes. This leads to the specification of the features of the further working steps, essentially the slicing and building procedures. The surface quality, and also the accuracy, are closely related to the arrangements for slicing and building.

That means that the model has to be oriented in three-dimensional space for optimum building.

A support structure is then added to hold the part in place while the layers are being built, to maintain the layer registration and the structural integrity of the part /10/. The support structure attaches the part to the elevator platform, and will be finally removed after the part is manufactured. The correct construction of the support structure is a problem area, especially for parts with overhanging features and difficult element shapes.

The best arrangements of the part-specific support require some experience. For relatively level parts, the usual vertical (z-direction) dimension for the support structure will be chosen at about 7mm. These structures also support the surface of the part so that they will not warp, sag, or parachute as the elevator moves up and down in the resin. Supports prevent the part from falling over as the wiper arm moves across and compensate for any angular misalignment between the elevator and the surface of the fluid /10/.

If the required size of the object is larger than the available volume in the SFM equipment, there are two possibilities:

- a change of the measurement (if the object will only be used for viewing by the designer, not as a model or functional part)
- subdivision of the entire volume into suitable smaller sections (it is no problem to fit and glue the sections together after the building procedure, using the same resin).

At the preparation phase, some part-specific requirements have to be considered. As an example, since it is necessary to take specific requirements of the objects into consideration, the influence of closed volumes can be considered.

Following the principle of levelling after every layer is hardened, the fluid will be moved into the contrary directions. That leads to sloping planes, which cannot be avoided during the building procedure. This general problem can be solved only in the following ways:

- by avoiding closed volumes, making other arrangements of the part, or
- including extra holes, which can be closed after the building procedure is completed.

Some more specific requirements, which cannot be explained here, have to be taken into consideration. Another general problem is the choice of tolerances in connection with the approximation of a circle by facets (see Fig. 4.2).

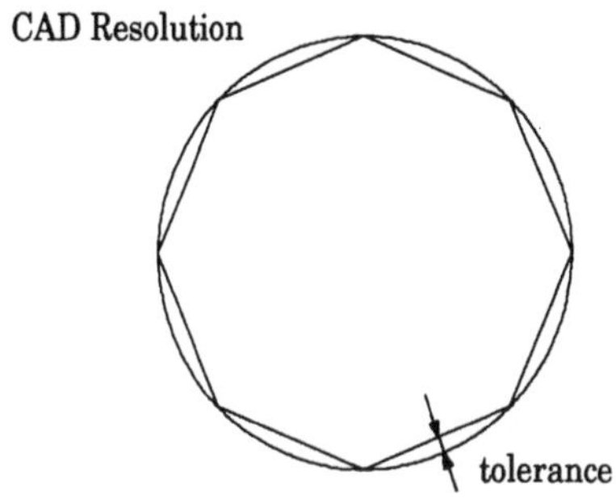

Fig. 4.2. Faced approximation of a circle

The larger the number of polygons, the more closely is the surface approximated. This results in a smoother curve on the finished part. When faceted data is used to drive the SLA, the approximations are reflected in the final part. To minimize this effect, the number of polygons used should be increased. However, as the number of polygons or facets increases, so does the slicing time and resultant data file size.

As a rule, the lowest acceptable CAD resolution should be used. For large holes it is useful to make a comparison with a suitable cutting procedure. The minimum recommended wall thickness is usually 0.020 inches (0.50 mm). Thinner walls are achievable under special circumstances, with the absolute minimum wall thickness set by the laser beam diameter (approximately 0.010 inches (0.25 mm)).

4.1.3 Stereolithography procedure : layer solidification

The hardening of polymeric fluids is based on the following effect. If the material is exposed to a laser or UV beam, the fluid is immediately hardened by low energy absorption. For better understanding of this procedure, the term voxel was introduced meaning a very small volume element (see Fig. 4.3).

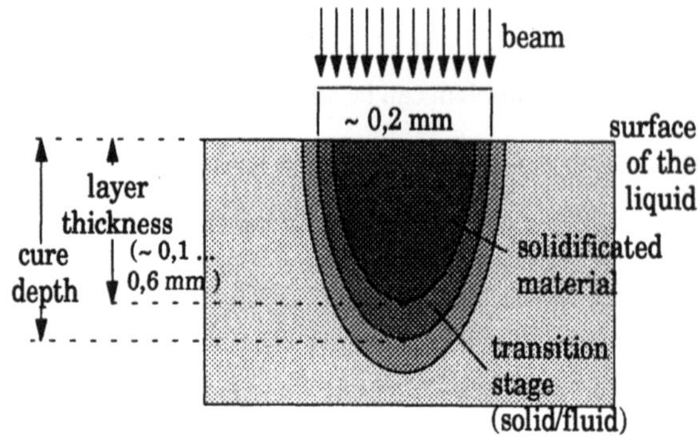

Fig. 4.3. Principle structure of the volume-unit voxel

The voxel size depends on the diameter of the laser beam and the hardening depths, usually in the range of 0.1 - 0.5 mm. For controlling the polymerization procedure, the laser beam is applied in micro second impulses. Between these impulses the beam is guided from one position to the next, in such a way, that a small overlap between the voxels will be guaranteed. This is achieved by using a step size than the voxel size (see Fig. 4.4).

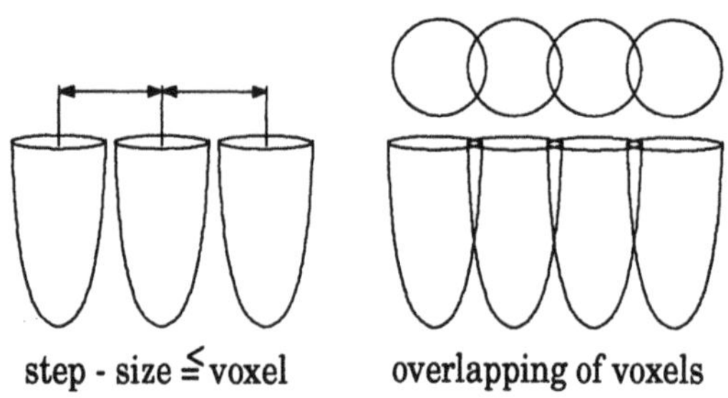

Fig. 4.4. Assortment of the voxel

A so-called hatch pattern is created. Different strategies have been developed for hatching. In this procedure a small hardened layer will be generated, in agreement with the sliced volume-element of the entire physical object. For building this entire object, overlapping of the voxels between the layers also has to be achieved, which means in the z-direction (see Fig. 4.5).

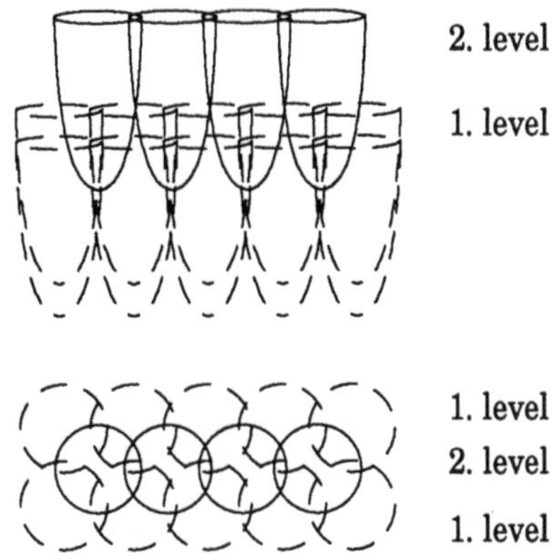

Fig. 4.5. Development of a single structure by voxels

This leads to multiple illumination of defined surface areas, with the consequence of additional hardening effects. The resulting problems could be uncontrolled shrinkage and deformations. For that reason, specific methods for guiding and controlling the laser beam are developed. Another typical feature of point-by-point layer solidification is the necessity for the additional building of support structures. The support structures have the following tasks.

- To separate the object from the platform. This ensures they are easily separated during postprocessing. The first object layer is of uniform thickness, even if the platform is warped. The resin can drain quickly through the platform to reduce the postprocessing time.
- To anchor the part firmly to the platform or to other part sections.
- To prevent distortion during the building process.

A specific and important factor related to Stereolithography processes is the necessity for support structures.

The support structures have to be prepared in connection with the CAD modelling or with the preparation of buildintg parameters, and in any case before the slicing procedure takes place.

In most of the equipment used the direction of the laser beam is from top to bottom, which means hardening the upper surface of the liquid (<a> in Fig. 4.6). Only the principle implemented by Mitsui (Japan) illuminates the lower liquid surface, through a glass plate (<b> in Fig. 4.6). This principle requires very liquid fluids and good possibilities for cleaning the glass. Based on this principle, specific features will now be explained.

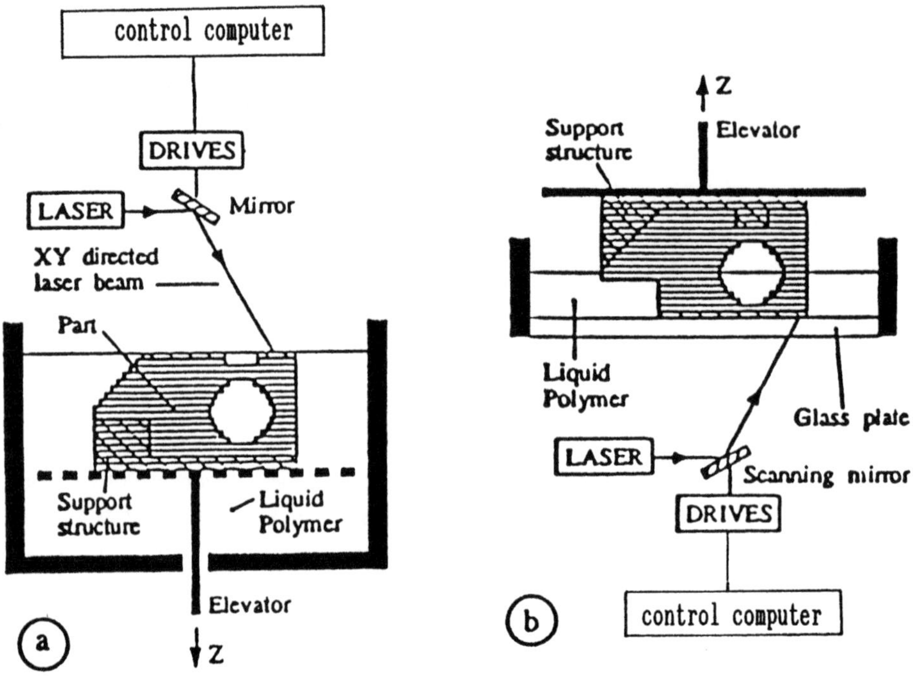

Fig. 4.6. Point by point scanning in different directions

4.1.4 Postprocessing

The main completion tasks after the solidification process are listed below.

Cleaning the so-called "green part", the part which is solidified, but not post-cured. Cleaning a Stereolithography part starts with its removal from the platform. Because the resins used, especially acrylate-based resins, can cause skin diseases, specific clothing should be worn and safety recommendations need to be considered.

The cleaning has to be done very carefully, because remainders of resin can lead to undesirable shrinkage or curling effects. For cleaning, ethyl alcohol is mainly used. TPM (Tripropylenglykol-Monomethylether) is most suitable for cleaning Stereolithography parts because it is

- safe (no explosion risk)
- approved for all kinds of resins
- reasonably priced, and
- no swelling is caused.

The following aspects have to be considered:

- resin saturates in TPM at about 12 %
- agitation of the parts is essential in the cleaning procedure
- the TPM and resin mixture must be treated as a hazardous waste.

Another popular method employs ultrasonic cleaning, wich penetrates into small crevices and openings.

Removing the supports

For this task it is useful to work with a special cupboard for chemical handling, including a ventilator. The handling of cleaning fluids requires, like the resin itself, the consideration of specific recommendations. The last essential step is the post-curing procedure with ultraviolet light at the Post-Curing Apparatus (PCA). Natural sunlight will also cure the part. After approximately 20 - 30 minutes, depending upon its thickness, the part is removed from the PCA. If necessary, the part can be sand-blasted, hand finished, and coated to provide a desirable surface finish.

4.1.5 Material requirements and properties

(see also Chapter 9)

The polymeric materials will usually be delivered in a canister, which avoids the influence of UV light. Additionally, it is useful to store this canister in a sheet metal cupboard for a guaranteed cool temperature. For the solidification procedure, such properties as viscosity and density are important. Furthermore, the tendencies for shrinkage, curling and some others have to be considered. Most important are the post-cured properties.

Some resins with particular properties, like

- tensile modulus
- tensile strength
- elongation to break
- impact resistance
- hardness

are available from different companies. Because of these sometimes extremely different characteristics it is necessary to make careful decision about the most suitable resin. The possible accuracies are closely related to the different material properties. Therefore the further development of material properties is an extremely important field of work. Some examples of suitable resins are described below.

4.2 Stereolithography (3D Systems)

The Stereolithography Apparatus SLA1 (3D Systems) was produced in 1989 as the first industrial application equipment, and has been continuously developed. Therefore this advanced principle will be described in fair detail.

4.2.1 Functional principle

The SLA principle is defined as follows: "Stereolithography is a three-dimensional printing process which produces copies of solid or surface models in plastic. The process uses a moving laser beam, directed by a computer, to print or draw cross-sections of the model onto the surface of a photocurable liquid plastic" (see Fig. 4.7). The solidification procedure is realized with a Helium Cadmium (HeCd) or Argon laser, in a layer-by-layer illumination from bottom to top. This means that the vertical elevator in the vat is controlled, step by step, from top to bottom. In line with the explanation of the operation of the NC technology, it is first necessary to define the coordinate system.

Coordinate system

There are three coordinate systems which are used in building stereolithographic parts /10/. They are listed in accordance with their sequence of use as follows:

- CAD system
- Slice system, and
- Mirror system (see Fig. 4.7)

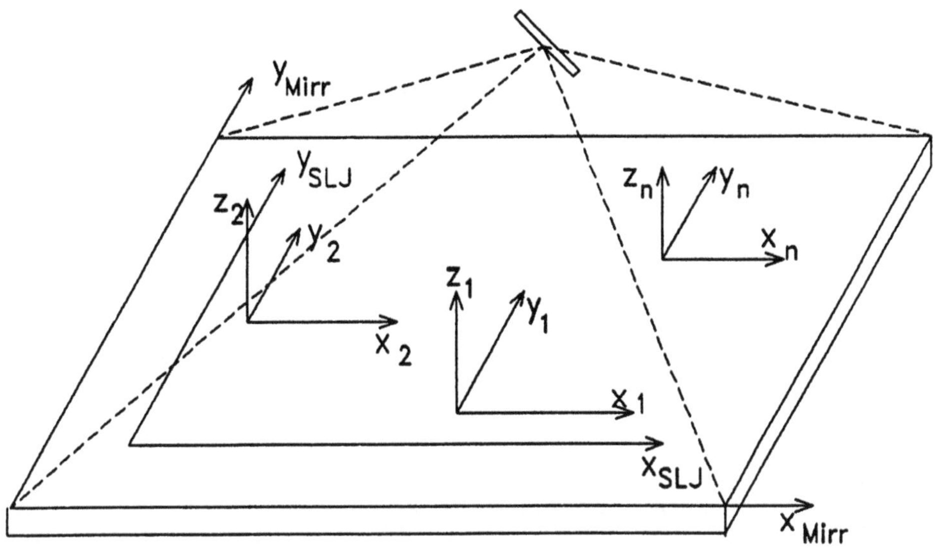

* $x_1\ y_1\ z_1$ several CAD-coordinate systems
. . .
$x_n\ y_n\ z_n$

* x_{SLJ} ;y_{SLJ} one slice-coordinate system for several CAD-objects (in different z-levels)

* x_{Mirr} ;y_{Mirr} one mirror-coordinate system for the entire scale

Fig. 4.7. Coordinate-systems for Stereolithography

In each system, different units of measurement are used (see Fig. 4.8).

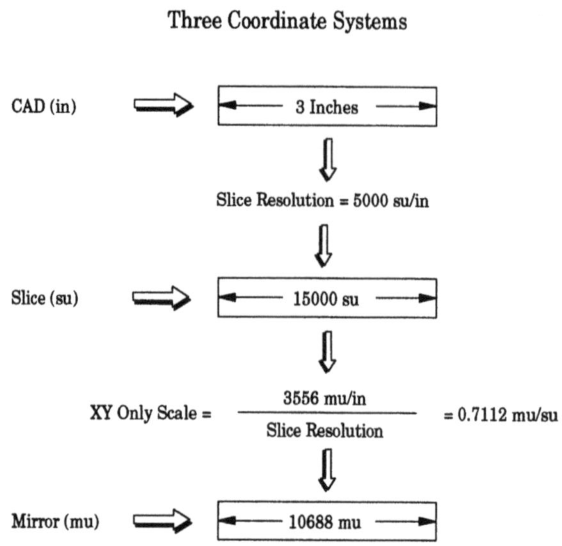

Fig. 4.8. Coordinate systems (measure)

The CAD System

CAD-system coordinates are established within the CAD database, and are used to orient the CAD model in three-dimensional space, relative to a global origin. The units within CAD systems are determined by the user, and can be either millimetres or inches /10/. In a CAD system, objects can be scaled, translated and rotated.

The sclice System

The slice system defines a three dimensional space that factilitates the cross-sectioning and vertical slicing of the model. A three-dimensional grid of slice units is superimposed on the CAD-space, with common origins. The user defines the size of the slice unit, which can range from 40 to 400 per millimeter (or 1000 to 10000 to the inch).

The mirror system /10/

Mirror space is a two-dimensional plane located at the surface of the liquid plastic. This system is used internally to position and move the laser beam by rotating a set of mirrors. The smallest distance within this coordinate system, known as a mirror unit, is 0.0071428 mm (1/140). Mirror space is normally transparent to the user. When a number is translated from one unit of measurement to another discrete one, the number must be rounded to fit in the other scale. During the Stereolithography process, this happens twice. The maximum error occurs when the number is rounded, up or down. The total tolerance can be calculated as follows:

- half slice unit (assuming the common slice resolution of 200/mm): 1 / 400
- half mirror unit: 1 / 280;
- Field of tolerance: ± 0.0061 mm

4.2.2 Hardware

The 3-D Systems Stereolithography-Apparatur is available in three types, in the following sizes:

SLA 190 : (7.5 x 7.5 x 9.8 inch = 190.5 x 190.5 x 248.9 mm
7.5 mW Helium-Cadmium UV laser)

SLA 250 : (10 x 10 x 10 x inch = 254 x 254 x 254 mm
16 mW Helium-Cadmium UV laser)

SLA 500 : (20 x 20 x 24 inch = 508 x 508 x 601.6 mm
200 mW Argon UV laser)

This is a relatively broad application range, not only by the possible measurements of the objects, but also by the performance of the laser power. The increased laser power allows higher speed and this leads to shorter process times. Because it is not intended to explain here all details of the different equipments, the main features of the SLA-250 will be described. The SLA-250 is a computer-controlled, electro-mechanical scanning laser system equipped with a recoater mechanism designed to print complex cross-sections on the surface of a vat of photopolymer /4.1/. The major components of the SLA-250 are shown in Fig. 4.9.

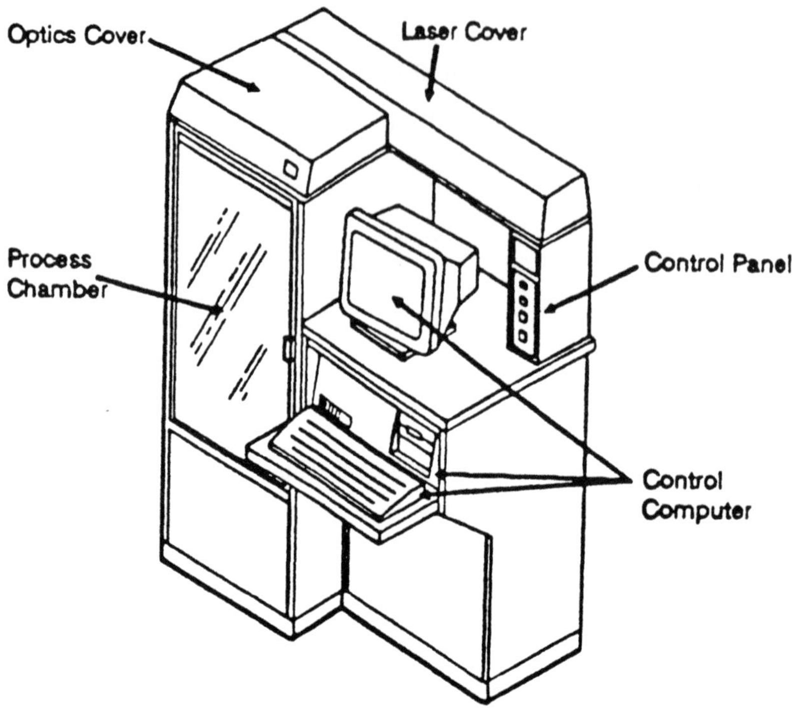

Fig. 4.9. Major components of SLA-250

Part of the basic equipment is the additional slice-computer. An essential and very sensible part of the SLA is the laser, together with the optical and scanning systems. This part is shown in Fig. 4.10.

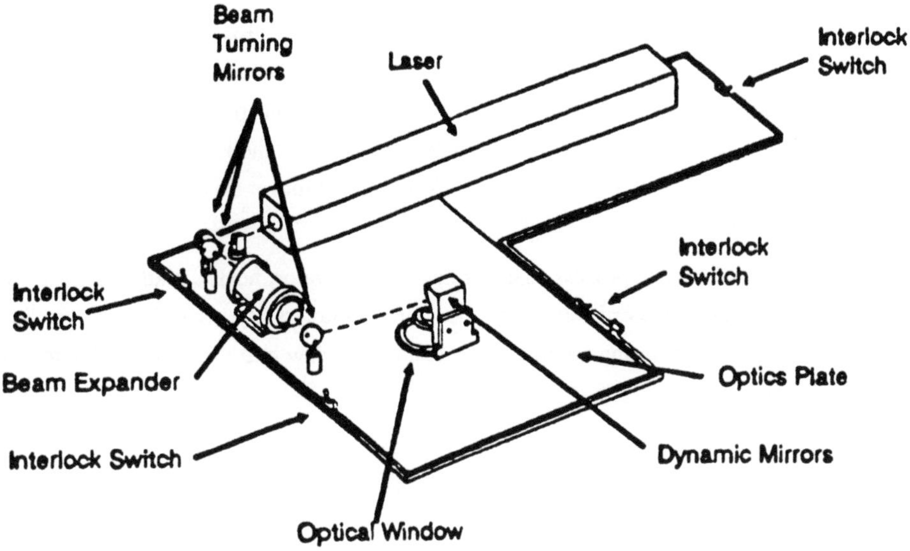

Fig. 4.10. Optical plate

The SLA-250 uses a helium-cadmium (HeCd) continuous, multi-mode ultraviolet laser. The initial laser beam power, 10 to 20 milliwatts, is at a wavelength of 325 nanometres. The optical system in the SLA-250 is composed of a shutter assembly, three beam-turning mirrors, a beam expander, and X-Y dynamic mirrors mounted on an optics plate above the process chamber. A 2-inch (50 mm) diameter precision optical window isolates the optical components from the chamber invironment while allowing the laser beam to enter the process chamber. One 20-degree, one 70-degree and one 90-degree beam-turning mirror, used to modify the beam direction, are coated for high reflectivity at 325 nanometers. The first two turning mirrors reflect the beam from the laser to the inlet aperture of the beam expander. The third mirror reflects the laser beam from the outlet beam expander aperture to the dynamic mirrors. A solenoid-actuated shutter operates to block the laser beam when an interlock switch is activated. Interlock switches monitor removal of the laser and optics covers and the opening of the process chamber door.

The beam expander unit expands the divergent laser beam to four times its diameter, then focuses it so that the beam converges to a small spot on the surface of the resin. This highly focused beam is desired for Stereolithography because it has a higher power density, which cures the resin more rapidly and more accurately. The scanning system consists of a two-mirror, two-axis galvanometer scan head and a mirror controller box. The high-speed dynamic mirrors deflect the laser beam, under computer control, to trace cross-sections by moving incremental distances (vectors) on the surface of the resin. The two mirrors are highly reflective at 325 nanometres and transmit most wavelengths. The full span of mirror movement is 65,534 mirror bits. A characteristic of the dynamic mirrors is that the laser beam will move 1 inch on the resin surface for every 3556 mirror units sent to the mirror drivers (1 mm for every 200 mirror units) /10/.

The environmentally-controlled process chamber houses an elevator, heater, platform, resin vat and beam profilers. The chamber itself is designed for operator safety, as well as to provide uniform operating conditions using a temperature control. A view into the process chamber is shown in Fig. 4.11.

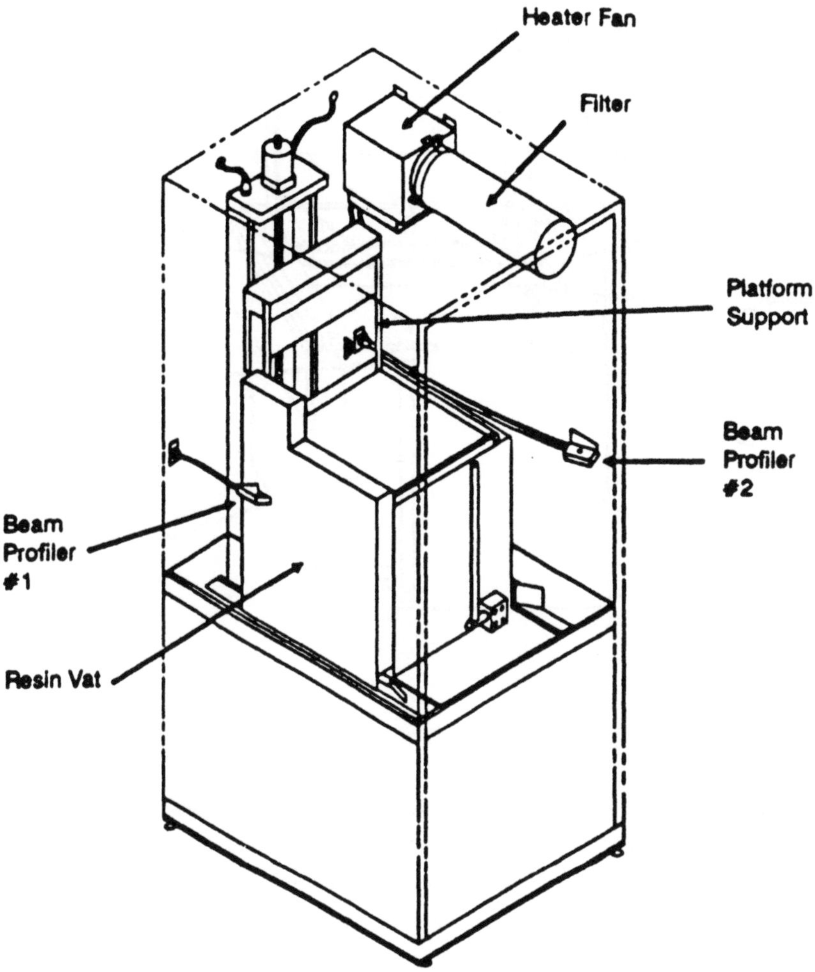

Fig. 4.11. SLA-250 process chamber

Before it is possible to start the solidification procedure, it is necessary to check the laser power and the laser profile. Two beam profilers, consisting of UV-sensitive detectors set beneath precision apertures, are mounted in the process chamber on both sides of the resin vat. The beam profilers measure laser beam power and profile, and are used in monitoring the dynamic mirror's drift. The laser power normally requested is 16 mW. The profile diameter can be displayed at the screen of the control computer; the normally used diameter is approximately 0.010 inches (0.25 mm). For a complete functional understanding, it is necessary to consider the essential elements around the vat (see Fig. 4.12).

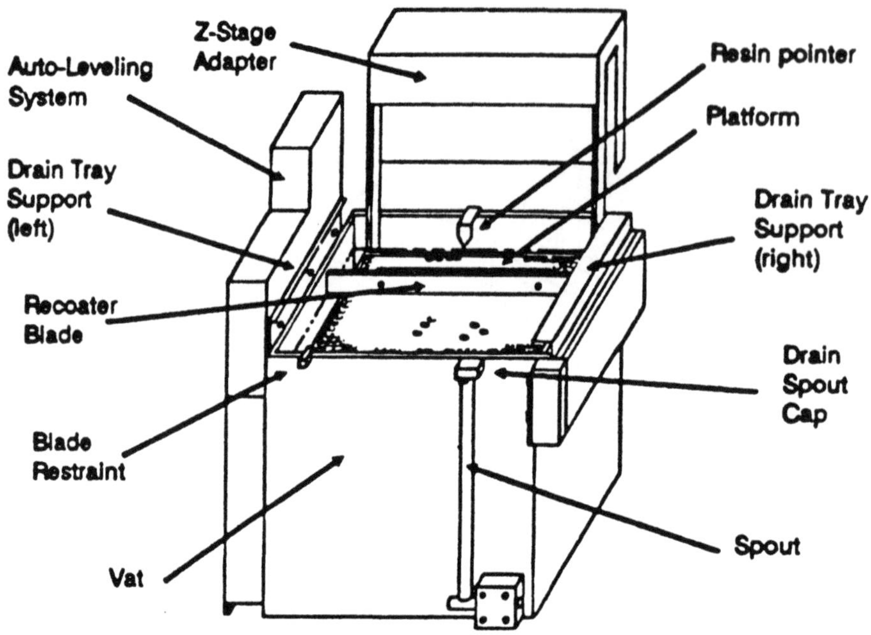

Fig. 4.12. SLA-250 vat and levelling system

Before the building procedure can start, the auto-levelling system has to measure whether enough resin is available. An automatic levelling mechanism, including a recoating system, is used to adjust the level of the liquid in the vat and maintain the correct amount of liquid on the top layer of the part. The control computer keeps the resin level constant during part building, to ensure proper layer thickness and recoater operation. Exact adjustment of the recoater blade is necessary for correct realization of the required layer thickness required. The elevator system consists of a power supply, a micro-stepping motor drive, and the elevator rail cable to which the platform support assembly is attached. The elevator raises and lowers, under computer control, the part concerned with the layer-by-layer- hardening process. The platform, attached to the elevator by platform supports, holds the part while it is being built.

The additional slice computer is used to generate cross-sections from the CAD design, for further processing on the control computer before part building. Two alternative computers are available as the slice computer: the NEC-UNIX based 386 computer or the SLP-20 (Silicon-Graphics UNIX-based Workstation). It is useful to link these above mentioned computers into an Ethernet network. Necessary software will be described below.

4.2.3 Process sequence

The entire process sequence is outlined in Fig. 4.13. For such complex process realization, many software packages are necessary. These software modules are summarized in Fig. 4.13.

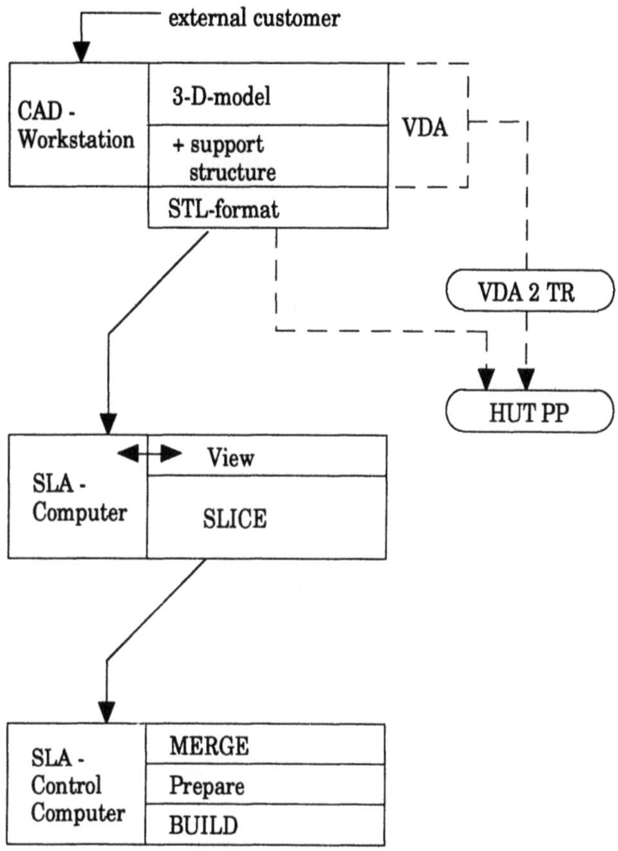

Fig. 4.13. Datatransfer and essential software moduls

4.2.3.1 CAD data transfer and CAD Interface

The CAD 3-D modelling is a very time-consuming task and normally has to be done externally by the customer, if for example the SLA is operating in a service office or in a large company. Moreover it is useful, in case of necessary corrections or modifications, to have the SLA equipment and also a suitable CAD workstation available. The design model and usually also the support data are converted to a Stereolithography format through a specific interface (see Chapter 7). This format is similar to a finite-element model consisting of triangular surface elements or patches /10/. The large amount of data can be transmitted from the external CAD workstation to the internal workstation or SLA equipment by various different means:

- network, for example Ethernet (if both equipments are in one company or laboratory)
- diskettes (for limited data volume of ~1,44 MByte at the 3½ inch diskette DS-HD; if the object requires more space, the data-volumes will be split automatically)
- magnetic tape (nearly unlimited capacity)
- e-mail, or internet, if available at both partners (limited capacity).

4.2.3.2 Software module VIEW

Definition: The viewing function is accessed from the Main Menu and displays Stereolithography files (.STL) and Slice files (.SLI) in graphical form. The viewing function is used primarily to study layer and part features, check Slice files and evaluate the need for variable layer thickness.

This is useful and necessary, as the system is highly dependent on human evaluation and interference /11/. If the computerized 3-D model and the sliced cross-section are carefully checked, the last stage of preparation at the SLA computer can follow. The SLA computer has to know how accurately the triangles define the original surfaces. If there are small holes (especially at crossing shapes) it has to be decided whether the object can be manufactured without problems. In the case of too large holes a warning is given if the object cannot be completely produced. Another essential is the checking of the correct direction of the normal-vectors. The normal-vectors determine the hollow space and the full volumes (see Fig. 4.14). The module VIEW uses triangles or vectors for data representation.

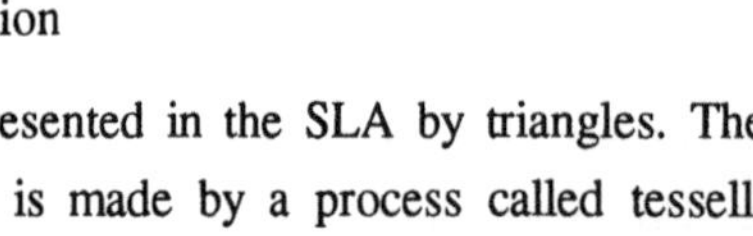

volume and hollow-space by normal-vectors

Fig. 4.14. Surface orientation

Surfaces or solids are represented in the SLA by triangles. The conversion from the CAD representation to triangles is made by a process called tessellation, which mathematically divides surfaces into planar facets or triangles. Many CAD and solid modelling systems represent surfaces of parts as facets for graphical display. Triangles, the simplest polygons, can closely approximate all surfaces. A rectangle can be represented by two triangles, while curved surfaces can be closely approximated using a large number of triangles. The SLA defines three types of triangles: flat, near-flat and steep. Based on the orientation of the CAD model, these triangle categories determine how the vectors are generated to represent the surfaces to build the part (see Fig.4.15). Vectors are very small lines which are traced by the laser to create the plastic part.

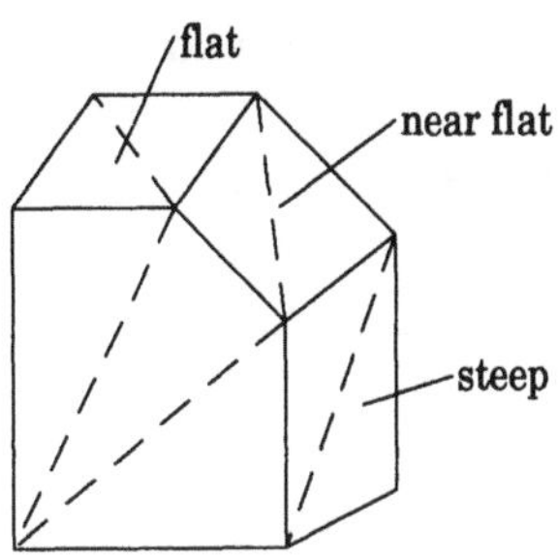

Fig. 4.15. Triangle classification

When triangles are sliced, three types of vectors are created to define the surface boundaries and internal structure on a layer-by-layer basis: layer borders, cross hatches and skin fill. The cross hatch vectors are an internal grid structure, created to strengthen walls and maintain structural integrity, while entrapping liquid in the area between layer borders. The spacing between cross hatches, and the type used, are selected by the user. The skin fill vectors define horizontal (top and bottom) surfaces. They are a series of closely spaced parallel vectors that form a skin when drawn by the laser (see Fig.4.16).

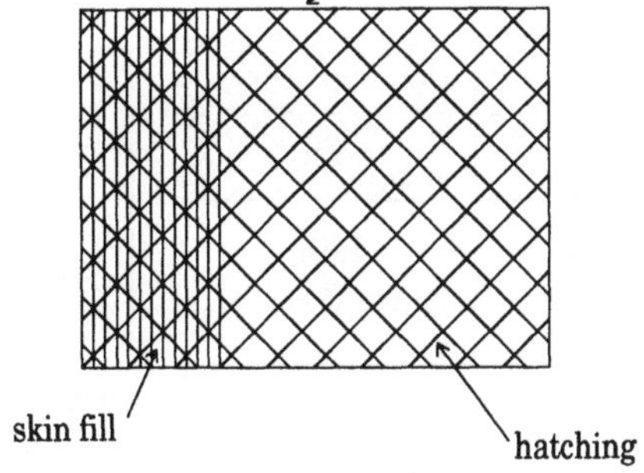

Fig. 4.16. Types of vectores

The module VIEW can also be used for all the other modules.

4.2.3.3 SLICE

Definition: Slice files consist of cross-sections layered one on top of another to create a three-dimensional object.

After checking and necessary correction of the modelled object, each file is sliced into cross-sections, typically of 0.12 to 0.50 mm /10/.

The user can select the layer thickness from consideration of the stair stepping effect (see Fig.4.17).

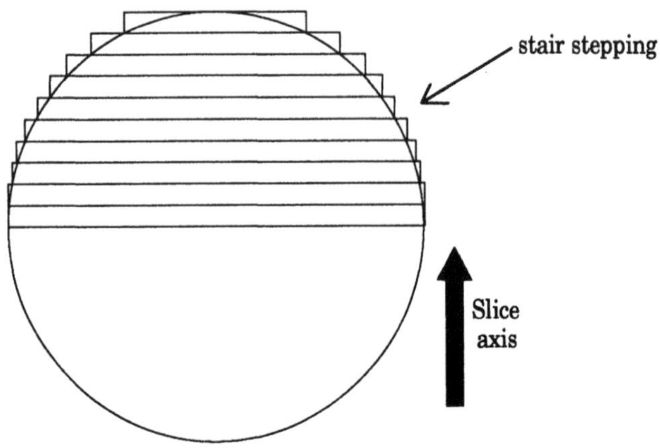

Fig. 4.17. Stair stepping of surfaces

Each slice is generated according to the requirement for necessary information of the final building of the object, and each holds information on contour, hatching, and skin /11/. The usual hatching distance is 1mm, which is related with good utilization of the laser energy. A graphical illustration of layer information is given in Fig. 4.18.

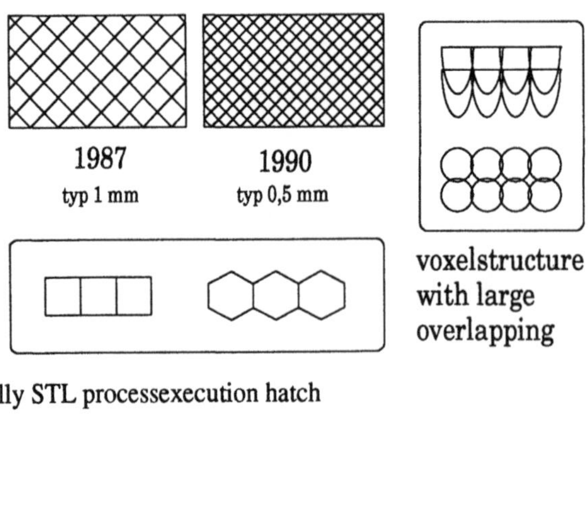

Fig. 4.18. Orginally STL processexecution hatch

The hatches can be chosen in 60°, 120°, x,y-direction or any desined combination. The hatches are an essential influence on the accuracy, especially in connection with shrinkage problems. For reducing this influence factor, some new software modules were developed by 3-D systems. The accuracy is mostly influenced by the process execution. Therefore different patented procedures are available. The traditional STL process execution Hatch is shown in Fig. 4.19.

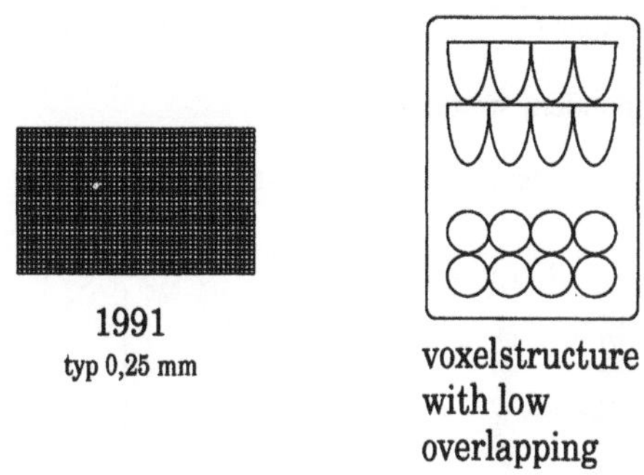

Fig. 4.19. New developed WEAVE

This procedure leads to a build-structure with large overlapping. The necessary post-curing of large volumes leads to the risk of distortions. As a result of research work, some advanced building methods aimed at error reduction have been developed /13/. This new STL process execution is shown in Fig. 4.20. Because it looks like a tightly woven piece of material, this cloth-like appearance of each layer surface accounts for the origin of the name "WEAVE" /13/. The result is a polymerization degree of ~95%, with better parameterization and relatively low internal stress.

The next development step leads to staggered process execution:

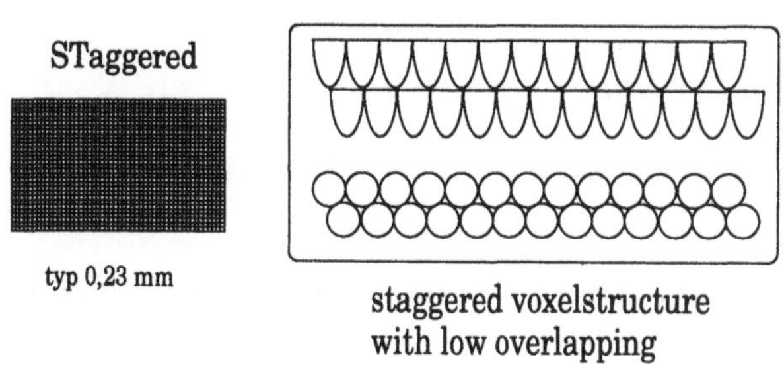

Fig. 4.20. STaggered principle

The degree of polymerization with the staggered structure is more than 98%. An additional aspect is the alternating process execution, linked with a geometry-dependent strategy:

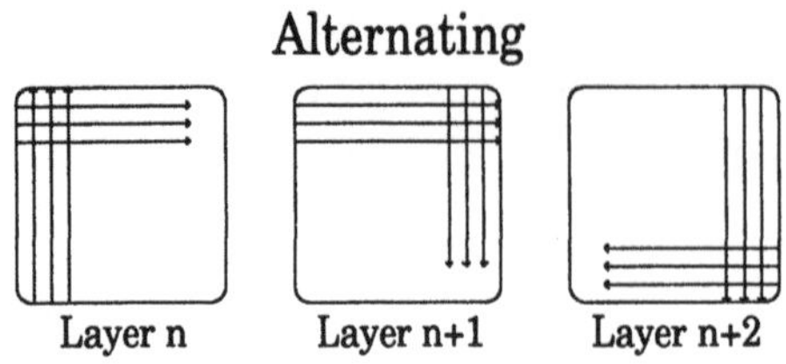

Fig. 4.21. Geometry - dependent strategy

A further reduction of stress can be realized by the so-called retracted method:

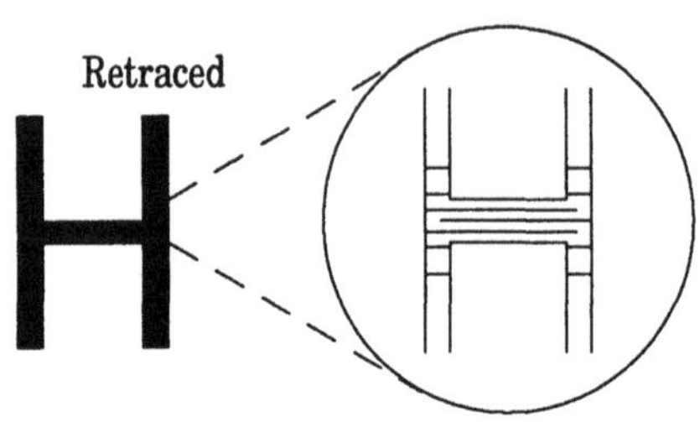

Fig. 4.22. Retraced hatch

By the term retracted it is to be understood that every second hatch is alternated connected with the borderline of the object. This leads also to better dimensional stability. Summarizing, the entire automated control software "STAR-WAVE" can be regarded as an important step in the direction of improved shrinkage uniformity and decreased post-cure distortion and dimensional instability. The overall part accuracy can be significantly improved.

Further working steps prepared by the module are the specification of the levelling parameters, elevator parameters and shrinkage parameters.

4.2.3.4 The MERGE module

This software module is installed on the SLA Control Computer. An additional Network Menu is therefore necessary.

Definition : The Network Menu allows remote operation of the Slice Computer from the Control Computer keyboard and file transfer through the LAN Ethernet between the Slice Computer and the Control Computer. The Network Menu is accessed through the Main Menu.

Definition : The Merge Menu, accessed from the Main Menu, combines all Slice files (support and object files) and generates SLA format data in the form of four files, respectively the Layer (.L), Vector (.V), Range (.R) and Parameter (.PRM) files.

- The Layer File (.L) defines the types of vector blocks in each layer.
- The Vector File (.V) contains vector data used by Build to draw each layer.
- The Range File (.R) contains user-specified ranges and parameters for Build. This file is modified using Prepare options to add part building parameters.
- The Parameter File (.PRM) is edited to control global part building. It is generated from the MACHINE.PRM file with viewport coordinates such that the part fill the screen and centred.

By means of this module, several SLI files can be merged into a group which can be used together in the further processes. For better utilization of the available space in the SLA vat, some different objects can be prepared for concurrent production. An example is demonstrated in Fig. 4.23 /10/.

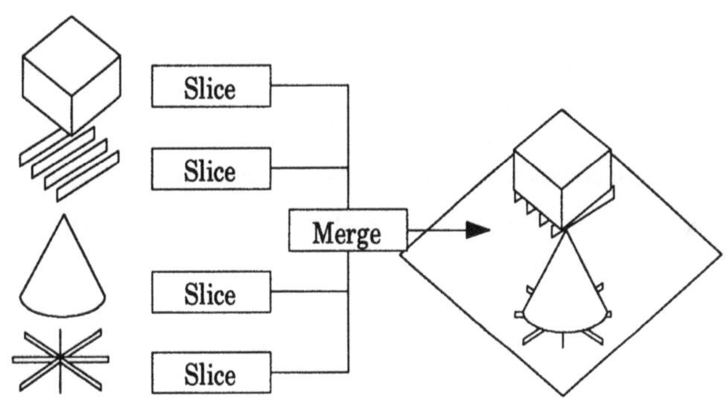

Fig. 4.23. Example of merging of two different parts to the platform

4.2.3.5 The module PREPARE

Definition: The Prepare Menu is accessed from the Main Menu. It allows the user to specify some building parameters such as cure depth, and essential recoating parameters for part building. The cure depth determines and degree of the speed the hardening of the resin by the installed laser power. The laser beam is guided in a step-by-step movement in accordance with the micro-second-impulses. This sets the depth of cured plastic that will be created. The cure depth is a result of the exposure to laser light. For a given laser power, this is controlled by two parameters: the step period and step size (see Fig. 4.24).

SCANNING SPEED VARIABLES

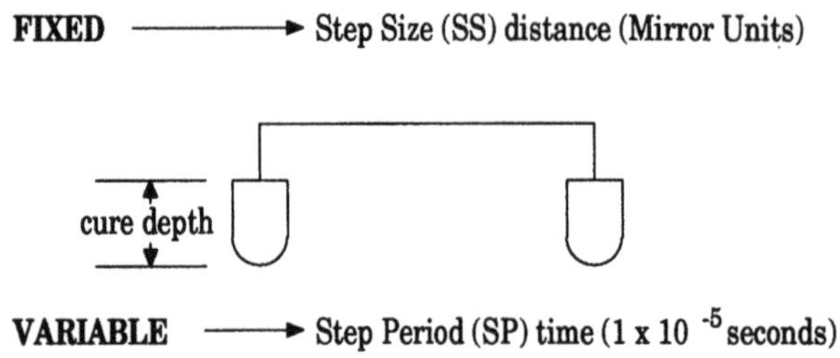

Fig. 4.24 Cure depths.

A step period (SP) is the length of time, in hundred thousands of a second, that the laser beam remains in one place plus the time required to move it from one place to another. The larger the SP, the slower the drawing speed and thus the greater the depth of cured plastic. The SP values for given cure depths are calculated from a working curve. The step size (SS) defines the size of the step, in mirror units, that the laser beam makes between forming bullets of cured plastic. Step sizes are calculated automatically within Prepare and generally do not require editing. The distance increment the laser beam moves is measured in units of 1/3556 inches (1/140 mm). Generally, the step size for borders and cross hatches is set to 2 (the minimum allowable). The step size for skin fill vectors is typically set to 16.

Another important term in connection with the preparation phase is the so-called over cure: the adhesion between layers is made by setting the cure depth to a specified amount in excess of the thickness of the layers (see Fig. 4.25).

LAYER OVERLAP

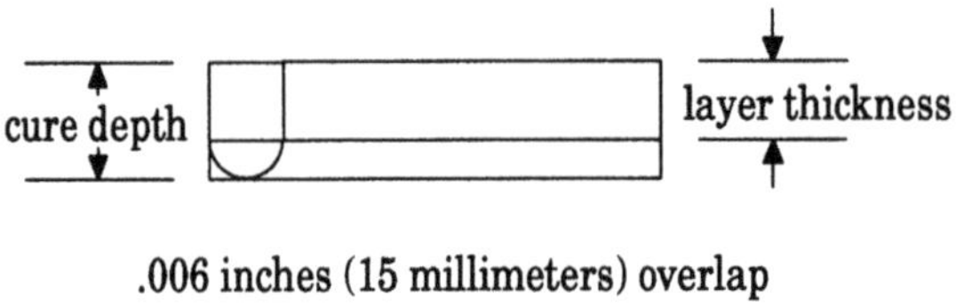

Fig. 4.25. Over cure

A Range is a vertical grouping of layers, created by the user, to specify a common part building parameter value. Ranges are defined by lower and upper dimensions, entered either in CAD or Slice units. One example for using a Range is specifying a recoating parameter for several layers. Ranges are often used to vary recoating parameters (see Fig. 4.26).

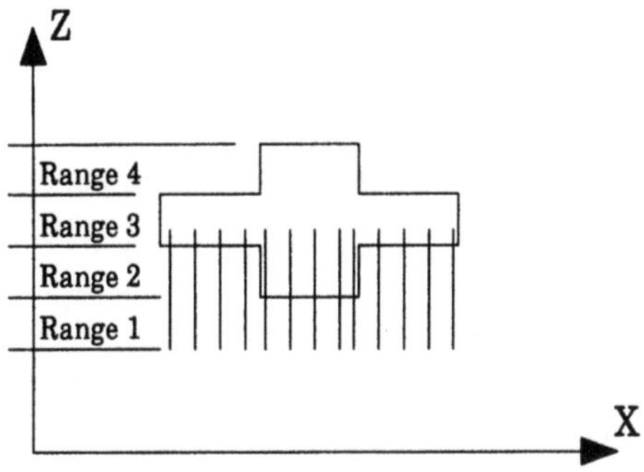

Fig. 4.26. Ranges

The module PREPARE has to be used also for determination of the elevator parameters: they can be entered into the Range file by using the Prepare Edit Recoating Parameters.

ZW This specifies the time, in seconds, for the liquid resin to level after recoating and prior to the start of the processing of data for the next layer. ZW ranges between 0 and 999, generally between 30 and 360 seconds with no sweeping. The default ZW during sweeping is 30 seconds.

ZV This parameter specifies the velocity of the elevator. A ZV value of 0.1 to 1.0 is recommended, or 0.05 to 0.1 for delicate areas. The default is 0.2

ZA This parameter sets the acceleration of the elevator. Again, a ZA value of 0.1 to 1.0 is recommended, with 0.05 to 0.1 for delicate areas. The default is again 0.2

Sweep parameters can be entered by using the Prepare Edit Recoating Parameters. A maximum of seven sweeps per layer is allowed. In general, the default of one sweep is used for part building. One of the most difficult decisions is the determination of the x-y and z-shrinkages. This allows for the part shrinkage which may occur, and may differ according to the resin being used. Another very specific influencing factor is the topology of the object. In general, determination of the shrinkage factor is currently the most sophisticated problem, requiring careful calculation and specific experience (see also Chapter 7).

4.2.3.6 The Module BUILD the model

Definition : Build controls the part building process by using the Vector (.V), Layer (.L), and Parameters (.PRM) files. The Build function reads and interprets the .V and .R files and sends the appropriate commands and parameters to the dynamic mirrors, recoating system, elevator driver, and vat. The Build functions include:

- Deflecting the laser beam to trace the vectors on the surface of the resin vat.
- Controlling the speed of laser drawing to ensure appropriate cure depths
- Controlling the dipping sequence, and recoating between layers
- Controlling the elevator position after Build

What the user sees being built while looking down into the vat is what Build displays in the viewing port on the screen. This "WYSIWYG" image (What You See Is What You Get) is in addition to a second window that displays the section of the vat that is being displayed at the viewingport. There is also the possibility of interference during the building of an object, for instance a break, jump or stop with a possible re-start.

O Utility Menu:

The Utilities Menu functions include applying power to individual components of the SLA, focusing and measuring the laser beam intensity, controlling the elevator platform, entering cure depth data, editing text files, checking the resin level, and moving the recoater and adjusting the blade.

4.2.3.7 Start the machine

This order begins the real part-building process on the SLA equipment. The processing time varies widely, depending on the specified layer thickness and the size of the part. On average, a small part will process about one inch (25.4 mm) of thickness per hour. One hour of continuous processing on the SLA-250 will cure four cubic inches /10/

4.2.4 Technological parameters

The technological parameters are determined during the slicing and preparation phase. Therefore there exists a specific recommendation paper which gives resin- specific instructions for making decisions.

The possible deviations, based on the shrinkage factor, could be in the working range between 0.1 - 0.9 mm. That means the accuracy is very much dependent on specific features of the determined object. Therefore, specific knowledge and careful preparation is requested. All parameters calculated are based on a speed of 381 mm/sec. For comparison, the speed of the SLA 500 amounts to 2540 mm/sec (based on the powerful Argonlaser).

The result of the building procedure can be influenced by several factors, such as:

- shrinkage
- swelling
- Cantilever Curl Distortion
- Post Cure Distortion
- Geometric Slab Distortion
- Creep

To avoid these failures, a careful process-realization is requested. Because of the relatively quick procedures, a repeated building process with changed parameters is possible.

Exact documentation of all essential part-specific process parameters seems necessary .

4.2.5 Application range

The typical application ranges of Stereolithography are:

- Prototypes for concept-models, offers, marketing
- form-fit functions for assembly tests
- models and patterns for vacuum-casting investment casting
- Patterns for metal-spraying; soft tooling
- manufacturing of electro-forms for EDM
- models, for stress-optical checking and NC shecking

4.2.6 Advantages and disadvantages

4.2.6.1 Advantages

The general advantages are represented by the following features

- the possibility of manufacturing parts which are impossible to produce conventionally in a single process
- for many application areas, the accuracy is sufficient
- the process computer provides good user support
- continuous unattended operation for 24 hours

4.2.6.2 Disadvantages

The development of the mechanical components seems not optimal, for the following reasons

- a sophisticated sequence of processes
- the laser quality is very variable, and the laser price relatively high
- necessary to have support structures
- relatively high expenditure for careful cleaning operations
- accuracy not in the range of mechanical part manufacturing.

4.3 Stereography - EOS

4.3.1 Functional principle

The core of the process of point-by-point layer solidification is the hardening of a liquid photopolymer by an ultraviolet laser beam. The description of the Stereolithography principle can be broken into its two main aspects

- consideration of the human factor by some appropiate hardware and software features
- supporting the process of "Reverse Engineering" by efficient coupling with rapid measurement equipment (EOSCAN)

The process of digitizing the actual model into the computer format should allow rapid corrections or changing of geometrical object data (see Fig. 4.27).

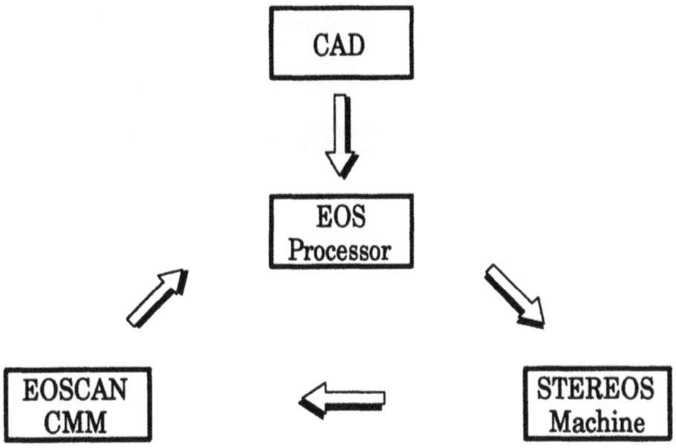

Fig. 4.27. Concept of advanced SFM-System

This concept is summarized in Fig. 4.27, which shows the essential elements for introducing a Stereolithography system into a company. At the top is the CAD system, which in many cases already exists. This is used to design and modify parts, and may also be used to generate tool paths for conventional manufacturing equipment (CAM). Standard data-transfer formats (such as STL, VDA-FS or IGES) are available, but problems can occur when trying to convert model files into different formats. A software interface is required to handle these different data inputs and enable the different parts of the system to communicate - this is labelled as the 'EOS Processor' in Fig 4.27. This processor also has to perform checks on the input data and optimize the process parameters (including part orientation, support structures, etc.).

4.3.2 Hardware and Software

The necessary components of the entire Stereography system are shown in Fig. 4.28.

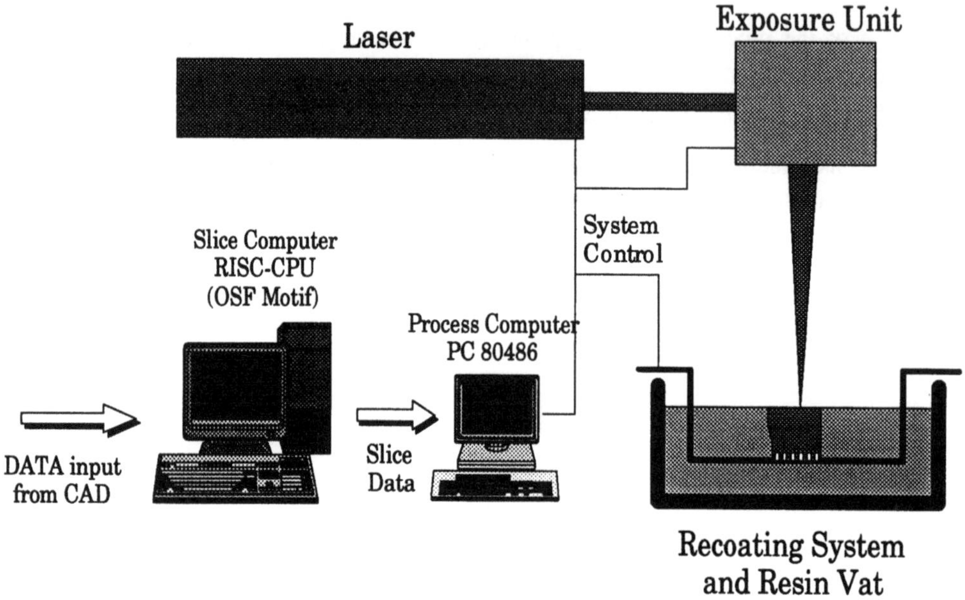

Fig. 4.28. Basic Stereography System

Two machine sizes are available, which can build models of up to 400 x 200 x 250 mm^3 and 600 x 600 x 400 mm^3 respectively. The model Stereos 400 is shown in Fig.4.29. Larger machines are also planned. Each machine can be supplied with interchangeable vats of 2 to 280 kg capacity, enabling small models to be built without requiring large quantities of resin. This is particularly useful when testing new resins, and can lead to significant cost savings.

Fig. 4.29. Stereos 600

The Laser

The machines can be operated with various ultra violet emitting lasers (typically Argon Ion) depending on power requirements. This gives the possibility of exchanging lasers or upgrading the system as required. The laser output is controlled from the computer, and for the lower power option (100mW Argon) no external cooling water is required.

Optics and Scanner

The laser beam is focused using a flat field lens, which enables a laser spot size of 0.1 - 0.15mm to be maintained across the entire vat. This produces a significant improvement over conventional focusing optics, as indicated in Figure 4.30. The two-dimensional galvanometer scanning system is also state-of-the-art, offering a vector speed of up to 10m/s at the liquid surface with a positioning accuracy of ±0.05mm. The laser beam position is automatically calibrated within the machine to ±50μm.

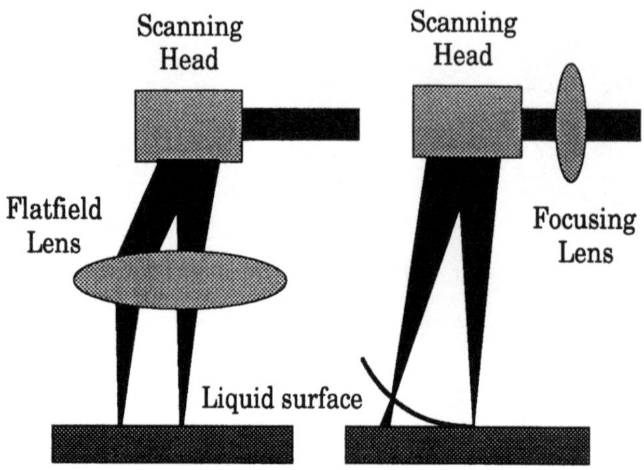

Fig. 4.30. Flat-field vs. conventional focusing lens

4.3.3 Process sequence

The first steps in the entire process chain will be realized by processor functions. The existing and planned functions of the processor are shown in Figure 4.31 as solid and broken boxes respectively. Software features include: interfacing with STL and CATIA for data input (VDA and IGES interfaces are planned, which can also be used for measurement data from a CMM); checking and repair of model descriptions; and automated support generation.

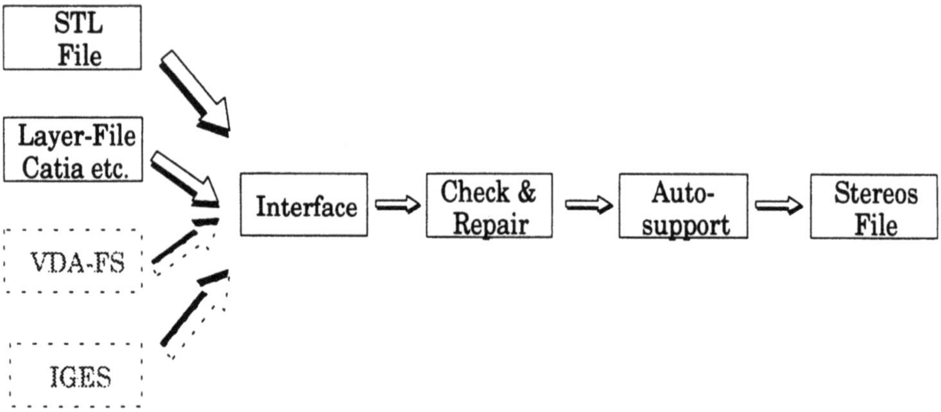

Fig. 4.31. EOS Processor functions

Process control in the STEREOS machine is performed by an IBM-compatible PC with 486 processor, operating under MS-Windows environment. This generates filling patterns on-line.

<u>1. Exposure unit</u>	
1.1 Laser : Argon-Ionen-Laser	
wave-length	350 nm
performance	50 mW, 100 mW or 300 mW
1.2 Laserbeam-guiding	
beam-extension-optic	6-times
focusing-optic (flat-field-lense)	
-Focal-length	800 mm
-Apertur	9 mm
-diameter of the focused beam (spot size)	0,15 mm
1.3 Laser-beam-deflection	
Speed of the laser beam	10 m/s
Positioning accuracy	± 0,05
<u>2. Hardening unit</u>	
2.1. Changeable Vat in two sizes	
- parts till 250 mm	70 l
- parts till 400 mm	~ 250 l (280 kg)
2.2 Positioning in Z-direction	
Vertical movement-accuracy	+ 0,05 mm
2.3 Level-regulation (automatized)	
accuracy	+ 0,02 mm
minimal layer-thickness	0,1 mm
<u>3. Process-Computer</u>	
Processor	80 486, 25 MHz
Memory	4 MB
<u>4. Workstation</u>	
CPU	32 bit RISC-Processor

Table 4.1. Technical Paramters EOS 400

Exposure vector control is controlled by a second PC-386 computer operating in parallel. The software is menu-driven and user-friendly, and process parameters are continuously monitored and displayed during part building.

4.3.4 Technological features

A specific feature to be noted is the **Recoating method**

A novel recoating method has been developed to improve performance. The new procedure reduces recoating time by eliminating dip delay, reduces bubble formation, and greatly improves the quality of parts containing closed volumes. With the rapidly changeable vat a range of different resins can be used, including all UV-sensitive resins compatible with argon ion lasers. Standard resins are SOMOS™3100, which is rigid but tough and machinable, and SOMOS™2100, which offers high flexibility. The interchangeable vats enable the resin to be changed rapidly without the need to clean the machine. Caused on the hardware-structure and the vertical laser beam an geometrical definition in one coordinat-system is possible, which allows an easy handling of the geometrical objects.

Another feature, and advantage, is the user-friendly menu for process control.The entire process preparation and realization is controlled by different software modules (see Fig. 4.32).

Tasks with high computing requirements by means of the workstation can be carried out.

Viewer allows a check of the input data, and also of the detailes Slicer carries out the slicing operation in determined layers and the calculation of the laser movements for any layer. The slicing-operation can be controlled by parameter files, in line with user requirements.

The command "Plazierer" starts the entire program and allows the following opertions: scaling, translation, mirroring, rotation and storage for all active objects.

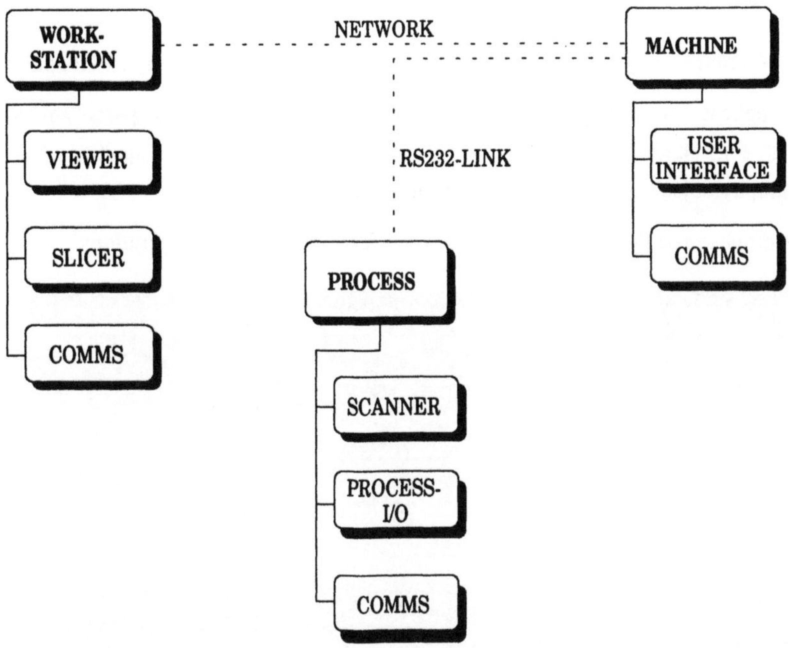

Fig. 4.32. Software moduls EOS

The software for process controlling and supervising achieves the object structuring and is the interface between the user and the control modules.

Essential tasks are:

- transfer of the SLI files to the process computer
- satisfying the following possible requirements:
  - sorting of the objects and determination of the illumination parameters
  - determination of the layer parameters
  - initialization and manual controlling of hardware units
  - start, interruption and continuation of the building procedure
- calibration and configuration of the global equipment parameters
- block building

4.3.5 Application ranges

The application ranges are similar to those of all other SFM processes (compare Chapter 8).

Traditional applications are related to CAD verification and functional test parts. An example is shown in Fig. 4.33. Because of the high accuracy and surface quality the EOS-Stereolithography objects are very well suited for manufacturing models.

It should be emphasized that all objects can be used for vacuum casting operations.

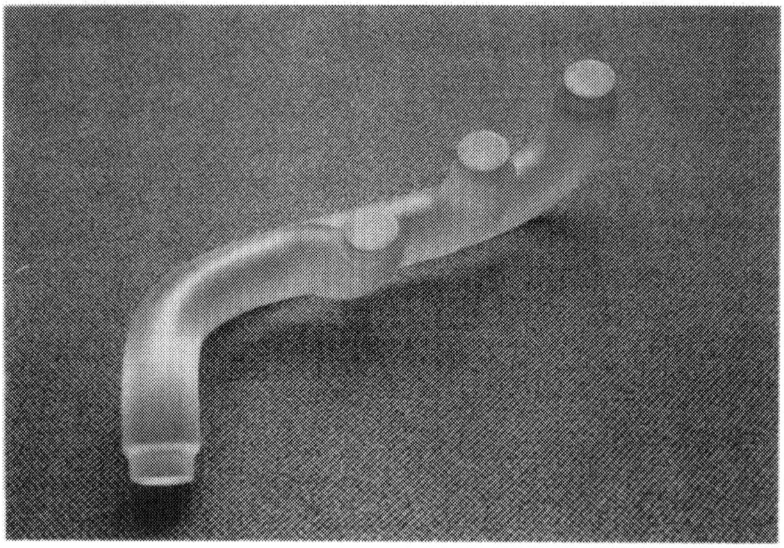

Fig. 4.33. Example for CAD-verification

The specific requirements for investment casting operations may be satisfied by models of high complexity, as shown in Fig. 4.34.

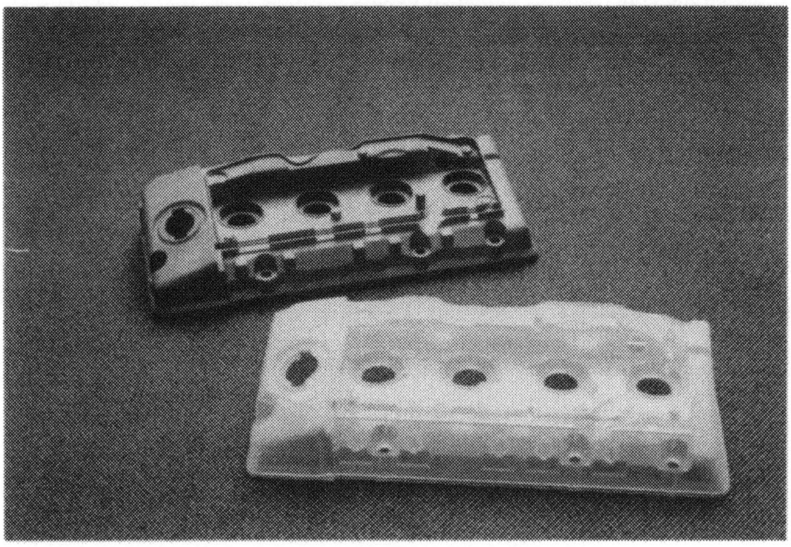

Fig. 4.34. Example for a functional testpart (investment casting)

4.3.6 Advantages and disadvantages

The remarkable advantages can be listed as follows:

- by using special changeable vats, the changing material is possible without problems
- the laser can be adapted to chosen requirements by continuous variation of intensity
- the hig performance of the laser beam combined with laser focusing by lat field lenses guarantees similar focus over the entire illuminated area
- the scanning speed of 10 m/s max is at the upper limit of all available Stereolithography systems
- because of the high speed and the very short switching times of the laser 10 - ... sec) an extremely accurate and quick process realization is possible. This allows an average illumination time of approximately 20 sec per layer
- unlike other Stereolithography procedures, a closed volume can be built without problems. This can be explained by the elastic wiper at the levelling equipment
- The quick scanning and recoating systems result in short building times. The geometrical object is always under the surface of the polymeric liquid, so delay times can be minimized

- based on the online generation of the hatch data, a drastical reduction of the data amount to approximately 5 % (in comparision with other Stereolithography procedures) is possible. By this reason the time for the preparation of the building procedure can be reduced in the same region
- stable operational accuracy of the entire equipment by the built-in calibration system

The disadvantages are general features of all Stereolithography system:

- restricted areas of application due to given material properties
- necessity of support structures
- labour requirements for postprocessing, especially cleaning.

4.4 Solid Object UV laser plotter (SOUP - CMET Japan)

4.4.1 Functional principle

In many respects this system is a modification of the Stereolithography principle, but it has some specific features. Whereas with Stereolithography the laser beam is controlled by a mirror system, the SOUP system uses an x-y scanner.

By using, this principle the laser beam is always directed vertically on to the resin surface. The laser power is controlled by a shutter/filter.

Both aspects contribute to high accuracy. Another specific feature is the use of epoxy resins with low shrinkage factors and good mechanical properties.

The principle is shown in Fig. 4.35.

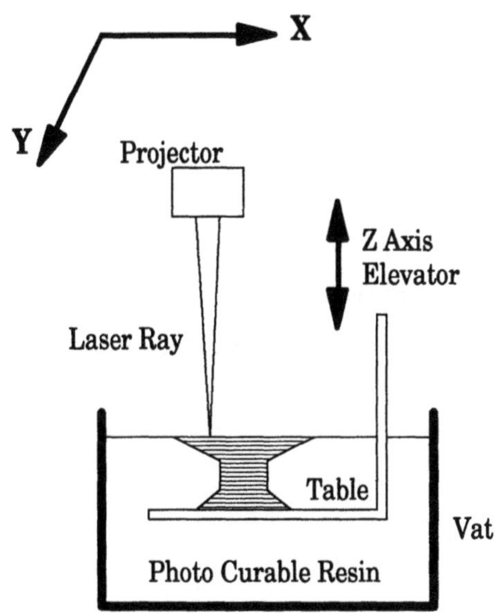

Fig. 4.35. Principle of Laser Lithography

4.4.2 Hardware Configuration

The entire SOUP system consists of a laser lithography unit, the SUN workstation (SUN4/40FC-8-P40) and the SOUP controller (see Fig. 4.36).

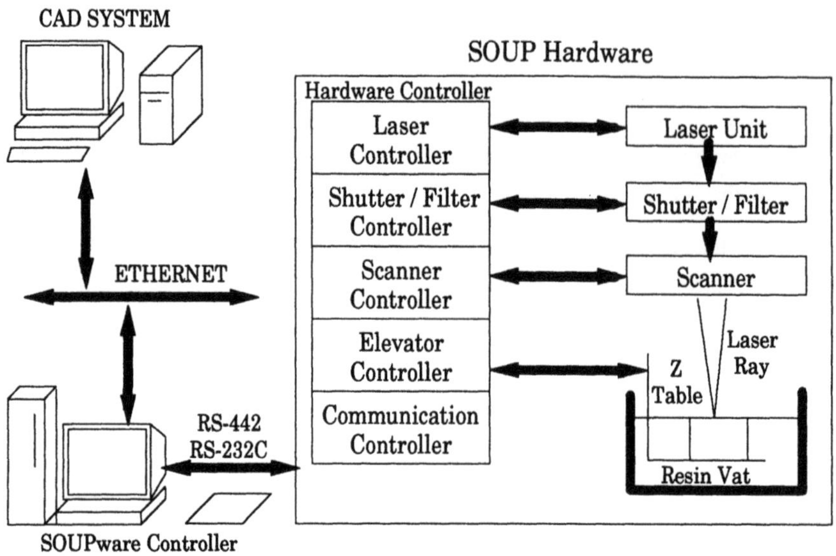

Fig. 4.36. Configuration of SOUP system

The CAD-system is linked via Ethernet to the SOUP controller.

There are several suitable CAD workstations. The controller workstations are SUN-3 series, SUN-4 series or SPARC station series.

The laser lithography units are delivered in different sizes . The SOUP 600 is shown in Fig. 4.37. Most remarkable is the high precision of +/- 0.05 mm, and the variable slice width of 0.05 - 0.03 mm.

Fig. 4.37 SOUP 600

4.4.3 Process sequence

The input data comprise the object description, as a surface or volume model, and must be translated into a SID (SOUP Interface Data) format. The following software functions are available for the process.

Convert

The Convert function generates the SOUP internal data format

- SLA Format (in agreement with the 3-D Systems standard)
- plane polygons
- Bezier surfaces

The interface to nearly all CAD/CAM systems is given via the SLA format.

Config

The manufacturing parameters are defined with the aid of this module:

- adjustable catching-tolerance for the closing of open contours
- interactive definition of support constructions
- automated construction of support bases
- generation of several modeles in one operation with different manufacturing parameters
- structure definition of every model possible up to 10 z-segments
- laser beam diameter adjustments independent of each other for other contours, internal structures and support constructions.

Execute

This module includes the programs necessary for the mathematical preparation of the data model and the manufacturing parameters in connection with the generation of NC data for the manufacturing process.

- direct generation of the NC data with simultaneous production
- generation of NC data in advance, with graphical simulation of the manufacturing process (determination of CAD mistakes)
- separated generation of contour data of every cutting plane for checking of the CAD data (displaying of open contours)
- offset function for mathematical compensation of the laser beam diameter

Side

This module includes a convenient editing facility for manipulation of the model data.

- function for automated positioning of a model
- mirroring, scaling, rotation, and translation of the model
- copying and adding of several models with automated positioning

On line

This function allows the checking of machining parameters directly at the workstation

- liquid level in the vat
- lasee performance
- positioning of the plotter head

Status

Gathers information about the current status of all processes which are initiated by the SOUP system

System parameter

Displaying of system parameters (access path)

4.4.4 Technological and economic features

Essential features of the SOUP system are:

- minimized waiting time of the equipment, because of the low resin viscosity
- no post-curing required
- interfaces to all normally used, CAD systems and also UDAFS available
- standardized CNC interface for data transfer to the SOUP equipment

4.4.5 Application range

The application range is the same as for all other point-to- point solidification systems.

4.4.6 Advantages and disadvantages

Advantages are :

- the application of resin with high viscosity
- the possibility of complete simulation of the building process
- the possibility of controlling the laser beam

- no post-curing required

Important disadvantages are:

- the general restrictions imposed by the applied materials
- the necessity for building support structures
- the mainly manually operated cleaning procedure

4.5 Solid Creation System SCS (D-MEC CAD)

4.5.1 Functional principle /18/

The functional principle is the same as for all other laser lithography methods.

4.5.2 Hardware configuration

The hardware configuration is, in general, also identical with other laser lithography equipment.

The solid creation system consists of the following:

- optical portion
- main mechanical part, including resin vat
- control computer, including a 32 bit workstation.

The SCS is available in three sizes, as shown in Table 4.2.

		Type	
Feature	ISC 1000	ISC 2000	ISC 3000
Laser (optical system)	He - Cd	Ar	Ar
drawing range (mm) movement range	240 x 240 150 mm or less	500 x 500 500 mm or less	1000 x 800
Spot size XY scanning speed	0.1 mm or less max. 250 mm/sec	0.15 mm or less max 3500 mm/lsec	0.3 mm or less max 3500 mm/sec

Feature	ISC 1000	Type ISC 2000	ISC 3000
Main body size	W1650 x D1400 x H1800 mm	W1650 x D1400 x H1800 mm	W1900 x D1550 x H2500 mm
cooling water	%	8.5 l/min	8.5 l/min
Electric power	AC 100V 10A	AC 200V (3phase) 45A	AC 200V (3phase) 45A

Table 4.2. Standard Specification for Solid Creator

All types operate with a galvanometer-mirror system. The maximum x-y scanning speed of 3500mm/sec is remarkable and can be considered a characteristic feature of the Solid Creation System. Other important factors can be listed as follow.

- in accordance with the material requirements, either a helium-cadmium laser (45mW) or an argon laser (500mW) can be used
- as a deflector, a high sweep speed galvanometer is used, which can generate both the raster scan and the vector scan
- lowering and stabilizing of viscosity can be done by temperature control of the resin

Features of further hardware components

1 AOM (Acoustic Optical Modulator)

Acoustic optical modultion enables high speed switching to be used.

2 High precision scanner

The new scanner system contributes to the high precision. Its precision is five times better than the old type of scanner. The performance of this new scanner is given by Table 4.2.

3 Variable field size control mechanism

A variable focal length mechanism is used, by which, it is possible to create small parts such as brings needed to scan in detail some area. In this case, it is possible to change the field size to 150mm x 150mm and 300mm x 300mm.

4 High precision squeegee

By using this new squeegee, it enabled to laminate on 10μm which was impossible before. Therefore, it is possible to apply for a ring model and an artificial tooth.

5 APC (Auto Power Control)

In order to achieve high precision, the SCS1000HD change the laser beam diameter dynamically while creating the model. Also it the laser power can be controlled automatically to keep the energy density constant. It is therefore, possible to create a high precision model without the warpage.

4.5.3 Process sequence

The process sequence is nearly identical with all other laser lithography processes. The entire process chain can be characterized in terms of the functions of the software

The main software functions are following:

- CAD Interface software
- Data Processing software
- Control software

The features of new software functions are described below.

1. CAD Interface software

The feature of the interface software is to provide various kinds of data formats:

- IGES including
  - Bezier surface
  - B-spline surface
  - Trimmed surface
- STL (especially related to PRO/ENGINEER AutoCad etc.)
- SC-FORM (especially related to CAT/A, DESIGNBASE, FRESDAM, UNICAD, SOLUTION3000)
- HP-GL (related to 3-D Measuring machine)

In addition, it enables processing of the nurbs surface, ruled surface and cylindrical surface data.

2. Data Processing software (EDIT functions)

In addition to present functions, new functions include the following:

- Indication of slicing direction.

  It is possible to indicate the slicing direction of the model on 3-D graphics view.

- Improvement of the support generating functions.

  It is possible to create various types of support shape (plinth shape, lattice shape in the empty body, etc.)

- Realization of high speed.

  This reduces the time to slice and thicken the model by about ten times compared with the previous functions.

3. Control software (Model Creation Functions)

In order to create a high precision model, the layers are controlled by setting parameters on each layer. The layer parameters concerned are the selection of scanning pattern, variable control of laser beam diameter and laser power, selection of vector/raster scanning, and so on.

4.5.4 Technological features

The general features are similar to those in other laser lithography procedures. The resin used is a methane acrylate, and is suitable as the coating material for optical fibres. This resin has a high sensitivity and a high speed of response, and so enables faster laser scanning.

4.5.5 Application range

The models made by SCS are used as design mock-ups, the functional model for checking functions combining with real parts, and the master model for manufacturing, and so on.

Some application examples are given in Table 3. The Sony Corporation is increasing its use of this system in the conceptal design field and in trial manufacturing. An essential feature of Japanese SFM technologies is the high level of medical application. One example is included in Table 4.3.

	Application
1) Toyota Motor Corp.	SCS´s model are sucessfully used for the development of the engine. The applied fields are : • conformation of the design • decision of the designing specification • stress analysis • fluid analysis of airflow
2) Sanyo Electric Co. Ltd.	SCS reduced 75% of the working model creation hours, so that effectiveness of 3D CAD is highly recognized.
3) Sumitomo Rubber Industries Ltd.	SCS was used to help designing the tread pattern of the tire and achieved high effectveness.
4) Medical Science Depty Kio University	The model of skull created from CT scanner´s data helped to succeed in the difficult operation of head

Table 4.3. Application examples of SCS

4.5.6 Advantages and disadvantages

These are closely similar to all other laser lithography processes (well mastered procedures; restrictions concerning material properties; considering toxic material behaviour, etc.)

4.6 Computer Operated Laser Active Modelling Machine (COLAMM)-Mitsui Engineering)

4.6.1 Functional principle

The COLAMM principle is also based on laser lithography.

The most important feature is, that the laser scanner is installed at the bottom, and the laser light is directed upwards through a glass plate.

4.6.2 Hardware Configuration

Suitable interfaces are the

- SLA format
- Contour line vector format for cross section data, or
- IGES.

As the postprocessor computer a sun sparc station is used. With the postprocessor, the cross-section data (vector data) can be generated. The main features of the model-making machine are:

- max model size : 300 x 300 mm
- laser type : HeCd (wavelength 325 nm)
- laser power : 40 mW
- laser beam diameter : 0.1 - 0.5 mm
- scanning speed : max 300 mm / sec
- model accuracy : ± 0.1 - ± 0.5 mm
- z-axis layer thickness : 0.1 - 0.4 mm

4.6.3 Process sequence

The general sequence is identical with the other laser lithography methods. The essential difference is the laser beam direction. That leads to the following process sequence at the model making equipment.

- first, the base is positioned slightly above the glass
- the laser beam performs the scanning controlled by the computer
- after laser scanning, the base is raised by the predetermined pitch
- then the laser beam performs the scanning again, and these processes are repeated, layer by layer

The main functions of the software are:

1) Displaying of 3-D CAD modeller data enlargement / reduction / rotation / view angle

2) Setting of operational function

- Cross section pattern (solid / grid / shell)
- Z-axis layer thickness (0.1 ~ 0.4 mm)
- Laser scanning speed
- Start layer/stop layer
- Preparing of table plate
- Offset

4.6.4 Technological features

Additional to the main characteristic, of the laser beam direction (see Fig. 4.6b), the essential features are those of the resin used, as follows:

□Photopolymer

- Principal element : Acrylate
- Specific gravity : liquid 1.1; solid (after curing) 1.2
- Colour : colourless (pale yellow)
- Viscosity : 980 cps
- Tensile strength : 470 kg/cm^2
- Bending strength : 740 kg/cm^2
- Shore D hardness : 84

(4) Photopolymer

The photopolymer consists of three elements:

- Oligomer
- Monomer
- Photoinitiator

The list below shows some characteristics of the photopolymer which is utilized in the COLAMM system.

(a) Hardening ability is very good

(b) Shrinkage is small

(c) Adhesive property between layer and layer is strong

(d) Non-irritating to skin

(e) Low viscosity

(f) No smell

At present, differing kinds of photopolymer are being studied so that the exact photopolymer for the customer's needs can be supplied.

5. Illumination of Planes

5.1 General principle

The general principle is again the hardening of photopolymeric materials. The essential difference from all methods explained earlier is that the surface is not illuminated in a point-to-point mode but by a complete lighting and hardening of the entire surface, using specially prepared masks.

Preparation of the masks uses the Xerox-method. Thus the layer which was generated from the CAD model during the slicing procedure will be put on a glass plate by an ionographical process.

In the working space, the mask will be positioned and a kind of flashlight derived from a UV light source will be triggered. At the illuminated places the photopolymer hardens, and the entire object can be manufactured, layer by layer.

The most important advantage of this method lies in its high accuracy because of the uniform hardening and avoidance of material tensions. The result is homogeneous bodies. The mask allows also very narrow tolerances for the outer borderlines of the parts. An important disadvantage of the method is the sophisticated equipment, which prohibits unmanned manufacturing procedures. This principle is currently used by:

- Solider (Cubital - Israel, USA, Germany)
- Light Sculpting (USA)

5.2 Solid Ground Curing Processes: SMP Solider 5600

5.2.1 Functional principle /19/

The Solider System 5600 uses the general mask principle described above in the following way.

Two cycles are required (see Fig. 5.1)

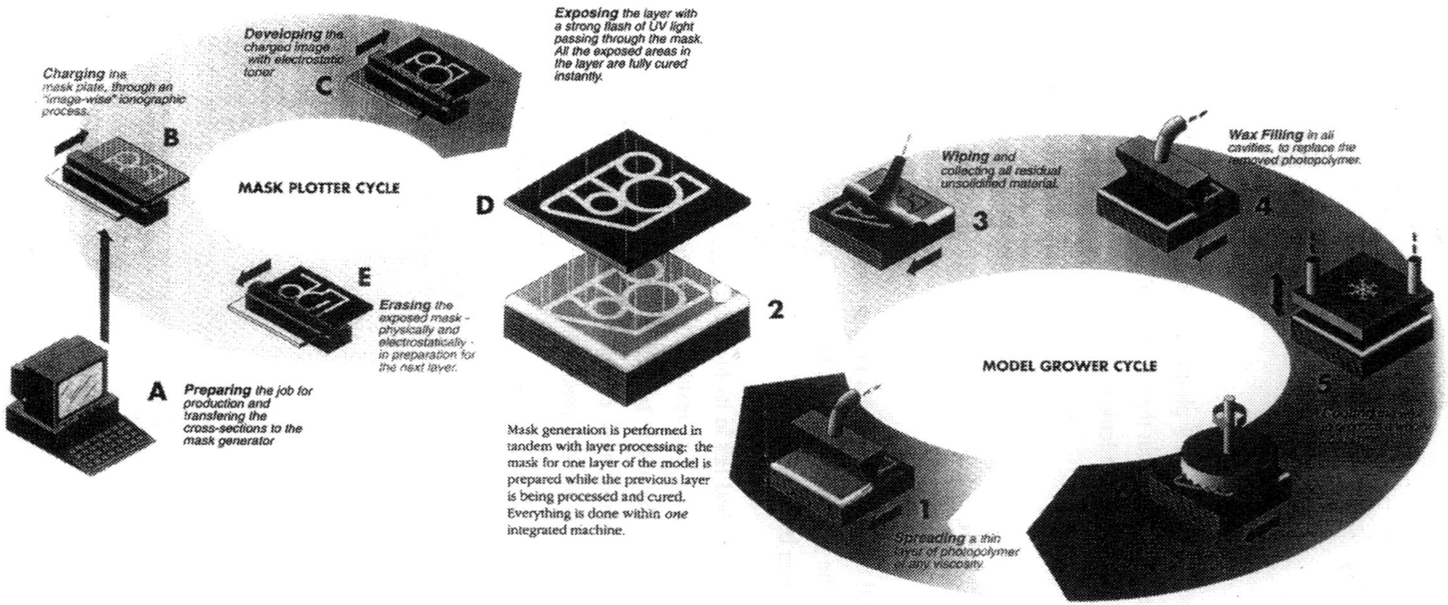

Fig. 5.1. Solider Cycle

1. Mask generator

On the left side of Fig. 5.1, the essential steps for mask preparation are shown:

- Preparation of the manufacturing procedure by transfer of the sliced CAD data to the mask generator (A)
- Down loading of the layer geometry to the control of the machine and loading the inverted shape of this shape to a glass plate by an ionographical process (B)
- Developing of the photopolymer by a powerful UV beam through the mask. All illuminated areas are hardened immediately (D)
- Delete of the mask; removal of the tuner and electrostatic loading (E)
- Repeat of the entire cycle

2. Model generation

- Applying of a thin layer of a photopolymer (1)
- Exposing of the photopolymer through the mask (2)
- Collecting up the non-hardened photopolymer (3)
- Providing a thin layer of wax, replacing the photopolymer (4)
- Cooling of the wax, producing a solid volume (5)
- Milling of the layer to guarantee the exact layer thickness (6)

Starting and repeating the same procedure cycle until the required object is completed.

5.2.2 Hardware

The Solider system's general layout requires dimensions of

$$L = 4.1\text{m} \qquad W = 1.7\text{m} \qquad H = 1.5\text{m}$$

and thus the most space of all known SFM equipments (see Fig. 5.2).

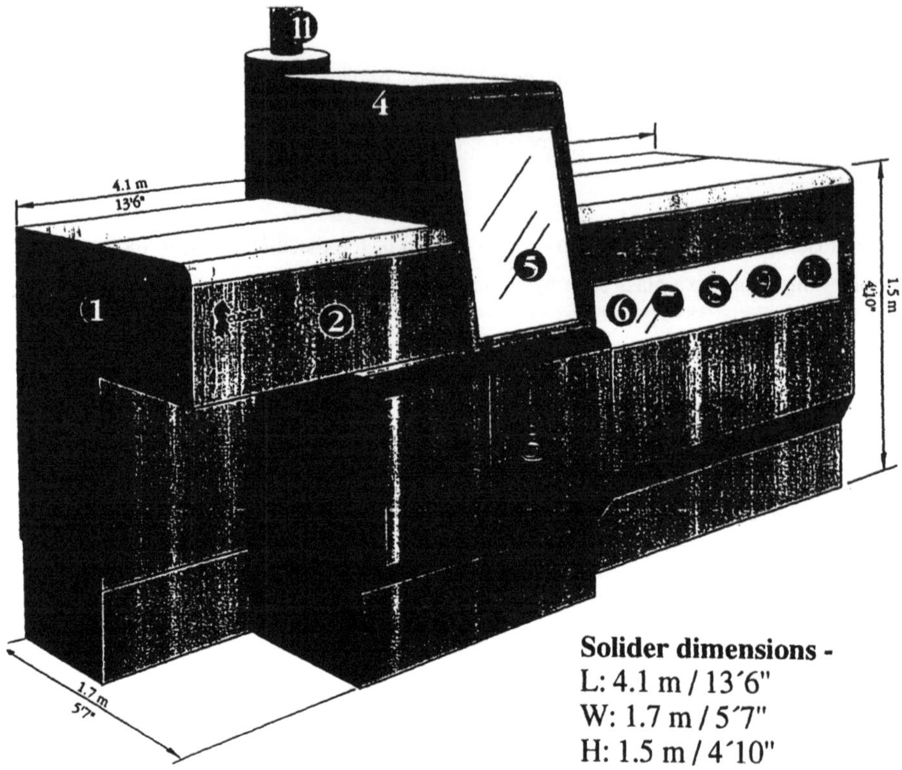

(1) MASK PLOTTER
(2) Optical MASK on a FLAT GLASS PLATE
(3) WORKPIECE CARRIAGE
(4) HIGH POWER UV LAMP
(5) EXPOSURE CELL
(6) RESIN APPLICATOR
(7) AERODYNAMIC WIPER
(8) MILLING HEAD
(9) WAX APPLICATOR
(10) COOLING PLATE
(11) AIR EXHAUST

Fig. 5.2. Solider System - general layout

This results from having the two units, for

- Mask plotting and
- Model growing

together in a closed area.

Because of the complete hardening of the object, no additional post-curing equipment is necessary. For removing the models from the wax block, an automatic dewaxing machine (washing machine) is available as an option.

5.2.3 Process sequence

The entire process starts with different possible CAD data input sources:

- Solid models
- Surface Models
- Wireframe models
- CT-scanning data (medical applications)
- Mathematical equations
- Chemical equations (molecular modeling)

Different CAD data from the customer can be used. Data from these sources are converted for production into Cubital's internal SoliFile format. Direct interfaces exist to SDRC-IDEAS (.UNV-files) and Pro-Engineer. Beyond these, the following standard input formats are accepted:

- .STL
- Binary .STL
- IGES
- VDA-FS 1.0
- VDA-FS 2.0
- .CFL (Cubital Facet List - a faceted representation alternatively to .STL)

Before the procedure can start there, a careful merge operation with different prepared objects is necessary. That is why the material consumption of the Solider procedure is relatively expensive.

Typically, several parts are built simultaneously during a single building cycle. The machine operator selects the parts to build and translates and rotates the data files representing them so that the group fits within a block of the smallest possible height. A further goal is to nest parts as closely together as possible, maximizing the number that can be fabricated at a time. After the parts are arranged, they are all sliced together to generate the cross-sections which control the mask plotter.

Once building has begun, the following sequence occurs for each layer of the part:

Step 1 The glass plate from the mask plotter moves right to station B, while charged by the ion source and developed with toner on its lower surface to form an exposure mask. The plate stops below the light source, at a precise location, and pauses.

Step 2 Simultaneously, the workpiece carriage transports the workpiece left to the station B, while picking up a layer of resin .006 - .008" thick from the resin applicator at station C. This resin layer is slightly thicker than the final layer thickness.

A stops at station B at a precise location and pauses. The surface of the resin is then a small distance below the toner layer of the mask.

Step 3 A shutter opens and light from a high-power 2000 W UV laser emerges through the plate, passing through transparent areas of the mask and curing corresponding areas of the resin layer below. The shutter is opened for about 4 seconds.

Step 4 The glass plate returns to the mask plotter, where it is cleaned of toner and prepared for re-use.

Step 5 Simultaneously, the workpiece moves past station C to Station D, where a stream of air lifts unexposed liquid resin towards a vacuum nozzle, causing most of it to be collected.

Step 6 The workpiece moves further to the right and passes a second flood lamp. The second exposure of the layer with UV-light increases the material strenght.

Step 7 As the workpiece moves under the wax applicator at station E, molten wax is spread over the entire layer to a thickness of about .008" (.2mm), covering the solified resin and filling in the regions from which liquid resin was removed.

Step 8 The workpiece then moves to station F and the wax is throught into contact with a flat cooling plate. The wax remains in contact with the plate until it is firm enough to be machined in the next step.

Step 9 The wax is separated from the cooling plate and then moves right, past the spinning fly cutter at station G. All the wax overlapping cured resin - as well as a small thickness of the resin itself - is removed, and a flat uniform surface of wax and polymer is produced by a cutting operation, with the scraps being collected by a vacuum cleaner.

Step 10 The carriage then drops down slightly in preparation for the next layer and returns to station C, to begin the cycle over again. After all layers have been formed, the workpiece is removed from the carriage and placed in an oven to melt the wax and release the parts trapped inside. Because the wax is also water-soluble, final cleaning is usually done with a warm water bath. Tolls and brushes may also be used to help remove wax from narrow passages.

The wax can also be melted away in a microwave oven.

5.2.4 Technological and economic features

One special feature of the solid Ground Curing procedure concerns the mask technology used and the accuracy achieved in the range of 0.1% of the part dimension. Factors affecting the accuracy include:

light diffusion in the glass plate, whose 5 mm thickness leads to double refraction. Furthermore, the distance of 3 mm between the glass plate and polymer upper surface also leads to light dispersion. In view of these and other factors, the offered accuracy of O.1 % is very high. Another specific feature - and if not treated carefully an important disadvantage - is the requirement to utilize the available vat volume as much as possible. The following calculation gives an idea of the cost.

The volume of the usable workspace is 500 x 350 x 500 mm. For a model with 150 μm layer thickness 220 μm of liquid resin are spread over the whole area. A full "job" on the machine consists of 3300 layers. The resin consumption - without taking into consideration the recycling - for this is 3300 x 0.022 cm x 55 cm x 35 cm ~ 127 liters. The cost of the resin for a full block of this volume is DM 12.700,- (DM 100.-/liter). With a "packing-factor" of 0.5, which is a good average estimation, only half of the resin will be used for building model volume. However, the other half is not waste material, but is collected for recycling by the producer. Today, 70% of the purchasing price of the original material are reimbursed for material that goes back to recycling.

5.2.5 Application range

The application range of the Solid Group Curing technique is much the same as for the laser lithography systems.

Additionally, a most interesting application range is given by the opportunity to manufacture small notches of O.1mm. That allows one to create the model's moveable parts or small mechanisms, for example gears in a gearbox. All kinds of irregular and complicated parts are suitable for this manufacturing process.There are no restrictions concerning deep holes, large undercuts, deep notches (bearing-space), thin walls and so on. That means the SGC techniques allow manufacturing of most complicated parts and mechanisms (see Fig. 5.3).

Fig. 5.3. Application examples Cubital

5.2.6 Advantages and disadvantages

In agreement with the survey by Levi /5.2/, the following advantages can be emphasized.

- The shrinkage effect is minimized, due to the full curing of every layer
- A high surface quality can be achieved
- No final curing in a special oven is needed
- No support structure is needed, the wax supports the model in all directions

- The structural strength and stability of the models are higher and they are much less brittle. This is due to the curing process that minimizes the development of internal stresses in the structure.
- Any geometrical shape can be made without limitation
- High throughput is achieved due to the three-dimensional nesting of models within the wax
- No hazardous odours are generated, the resin stays in the liquid state for a very short time, and the uncured liquid is wiped off immediately. Thus safety is considerably higher
- By altering the process slightly, wax models can be made and be used in the lost-wax casting process
- If there occurred any mistake during the building procedure it is possible to make corrections

Disadvantages

- The resin consumption of the machine is a constant, independent of the shape or number of parts but proportional to the number of layers
- The cleaning of wax from the models is relatively complicated and time-consuming (1 - 2 h)
- The Solider equipment is complicated, and therefore skilled staff are necessary. Unattended operation is not possible
- Because this is the most expensive of all SFM techniques, the economic justification is not so easy

5.3 Light Sculpting (LSI)

5.3.1 Principle /20/

The principle is similar to the Solider 5600 procedure. An electronically rewriteable liquid crystal display (LCD) mask is used. The photopolymer is then exposed by radiation through these masks. This principle avoids the problem of hand generating new masks for each layer.

5.3.2 Hardware

The hardware is available in 3 sizes.

1. LSI Photosolidification System, (Model LSI-0609 MA),

   (152mm x 152mm x 228mm)

   6 in. x 6 in. maximum part size

2. LSI Photosolidification System, (Model LSI-1212MA),

   (305mm x 305mm x 305mm)

   12 in. x 12 in.x 12 in. maximum part size

3. LSI Photosolidification System, (Model LSI-2224MA),

   (559mm x 559mm x 610mm)

   22 in. x 22 in. x 24 in. maximum part size

Essential technical parameters are given in Table 5.1

Forming Time, Maximum	40 seconds per slice of any size
Part Size, Maximum	6 inches x 6 inches x 9 inches 152 x 152 x 228 "X" by "Y" by "Z"
Slice Thickness	0.0127mm (0.0005 inches)
Slice Formin Accuracy	• X-Y Resolution: 0.003175mm 0.0254mm (0.000125 in. to 0.001 in.#) • X-Y Accuracy: 0.03175mm 0.0635mm (0.00125 in. to 0.0025 in.) • X-Y Alignment Accuracy: ± 0.001 in. • Z postion Accuracy : 0.002 in. • Thickness Resolution : 0.00254mm (below 0.0001 in.)
Radiation Source Data	Power Consumed : : below 200 Watts Service Life : : rated 10.000 hours Lamp Replacement cost : below $ 500.00
Cabinet Dimensions	26 in.(660mm) wide by 30 in.(762mm) by 75 in.(1905mm) high Other configurations available
Power requirements	115 VAC, 60 Hz, 5 A
Venting	Exhaust vent required

Table 5.1. Photosolidification Product Specification

5.3.3 Process sequence

Assuming a complete CAD model is available, it will automatically be already sliced, based upon the parameters of the part, and include the desired slice-to-slice dimensional change, slice thickness, etc., number of parts to be built simultaneously, scale in X, Y and Z, angle of rotation and other factors.

The general process sequence matches all the other commercialized systems, as is shown in the following.

1. Start with CAD data (separately generated) in one of the standard formats.
2. Then transfer the data into the 3-D solid modelling software.
3. Next, orient the part in the 6 x 6 x 9 in. work space for best fabrication, adding any necessary support structures.
4. After that, "slice" the model according to its unique geometry and tolerances using specific copyrighted software.
5. Then prepare a set of highly accurate, 2,470 dpi masks representing the part cross-sections.
6. Finally, fabricate the prototype using the patented Light Sculpting process

5.3.4 Technological features

The LSI procedure offers some interesting features.

- The CAD slice data is output to a floppy disk in PostScript format which can be utilized by a precision photoplotter to make the individual masks automatically.
- In the SLI attended system, all operations except mask changes are fully automated. Masks are placed in the machine one at a time by an operator when prompted by the control computer. Other operations such as shutter opening and closing, irradiation time, platform movements and the polymer deposition and levelling are automatic. The masks are sequentially numbered for operating convenience.

- Typically, a part may require 10 to 50 masks for every 100 layers built, but this ratio depends upon the part geometry. If the cross-section is constant in the "Z" direction, the same mask is kept in place as any number of identical layers are formed automatically.
- In the LSI system, a highly accurate mask is virtually in contact with the new layer of photopolymer, which optimizes accuracy and resolution. The extreme closeness dramatically reduces the amount of light energy required to solidify the exposed volume.
- During irradiation, the slice being formed is constrained from the top. This proprietary feature allows the LSI system to build many parts without the need for support structures, this reducing pre-engineering time and post-fabrication processing.
- When the part (or multiple parts) is completed, it is 100 % solidified and virtually completely cured. A short post-cure (12 - 15 minutes) is recommended to complete the chemical cross-linking of the material.
- If size permits, multiple parts or variations of the same part can be built in parallel without any production time penalty. The software easily duplicates the CAD data to allow building "nested" or positioned parts within one mask.

5.3.5 Application range

The application areas are similar to all other laser-based UV light procedures.

5.3.6 Advantages and disadvantages

The advantages are:

- high accuracy
- relatively high operational speed
- part building without support structures
- All horizontal surface are fabricated with the high-gloss, smooth, mirror-like finish which is inherent in the LSI process

Additionally, the part can be drilled, tapped or painted, or receive other processing if required. The cured material has a flexural modulus of 180,000 psi and a hardness of about 65 on the Shore "D" Scale.

The disadvantages are those common to all other procedures using photopolymeric materials. Additionally, the necessity for attended operation is an important disadvantage.

6. Solidification of Powder or Other Materials

6.0 General remarks

An important factor which leads to optimistic forecasts of the broad application of SFM-techniques in the near future is that some new materials are also suitable, as well as those currently used (mainly plastics) in powder, wire or laminate form. The originally developed principle was based on photopolymeric materials and opened up the new area of SFM technologies. These will undergo some improvements and further developments, so all of the application ranges described and certainly some new ones will remain important. But in principle other approaches offer possibilities of extending the kind of materials, which can be applied in other situations. The starting point with 3-D CAD model data is identical with the principle explained earlier.

6.1 Selective Laser Sintering - SLS-Sinterstation 2000

6.1.1 Functional principle /21/

At the beginning of the SLS process, a very thin layer of heat-fusible powder is deposited in a working space container and heated to just below its melting point. An initial cross-section of the object under fabrication is traced on the layer of powder by a laser. The temperature of the powder impacted by the laser beam is raised to the point of "sintering", forming a solid mass. As the process is repeated layers of powder are deposited and sintered until the object is complete.

The laser beam intensity is modulated to sinter the powder only in areas defined by the object's design geometry. In areas not sintered, powder remains loose and serves as a natural support for the next layer of powder and the object under fabrication.

6.1.2 Hardware und Software /22/

The SLS hardware consists of three main components. They are the Computer/Control Cabinet, the Atmosphere Control Unit, and the Process Chamber. The Computer/Control Cabinet houses the 486 UNIX based computer system, which is used to interpret the STL file and perform the slicing algorithm, and to monitor and control the entire SLS process.

The Atmosphere Control Unit houses the equipment for filtering gas recirculated from the process chamber, maintains a set temperature for the air flowing into the process chamber, and regulates the nitrogen atmosphere used in the unit.

The Process Chamber is used to house the laser and powder handling system (Fig. 6.1). The heat source for the sintering operation is provided by a CO_2 laser. Its beam is focused on the top layer of the part cylinder and directed by a system of optics and scanning mirrors. The part cylinder is capable of building parts up to 30.48cm (12.0 in.) in diameter and 38.10cm (15.0 in.) high. The part to be built in the SLS machine must be able to fit within this cylinder. Material is transferred from the powder cartridge feeding system to the part cylinder via a patented powder levelling roller. The hardware parameters are given by the specification in Table 6.1.

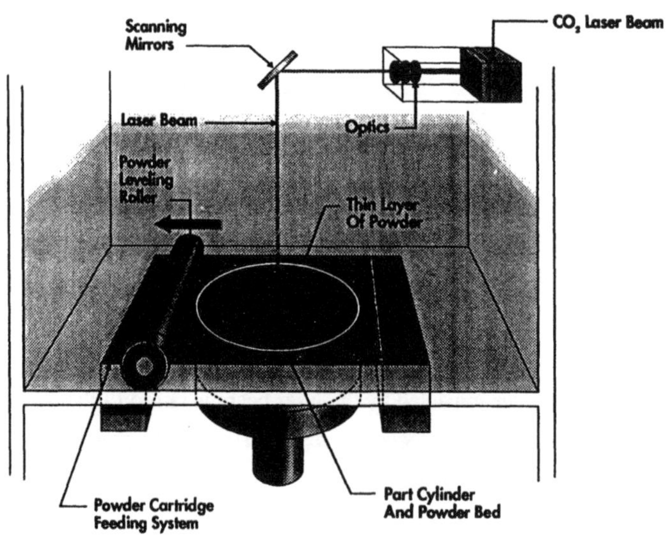

Fig. 6.1. Process chamber of the SLS Sinterstation 2000

Optical Parameters	Spot size :	approx. 0.020 in (l/c^2) - CO_2 Laser
Laser System	Type :	CO_2
	Wavelength :	10.6 μm
	Power on delivery :	50 Watts (maximum)
	Minimum lifetime power :	50 Watts
	Operating lifetime (minimum) :	2000 hrs.
Scanning System	Minimum step size :	0.000211 in (0.01mm)
	Spot location repeatability :	± 0.005 in (0.12mm)
	Scan speed :	30-40 in/s (1016mm/s)
Piston System	Part piston size :	13 in dia x 5 in deep
	Piston min. step size :	0.0002 in (0.005 mm)
	Piston accuracy per inch of working range :	± 0.0004 in (0.010 mm)
	Max. powder capacity :	2.1 cu ft (0.03 cu m)
Control System	80486 processor with 33 MHz clock	
	520 MB SCSI hard drive	
	80387 coprocessor	
	5.25" Floppy, 1.2 MB / 3.5" Floppy / ¼" Tape	
	16 MB memory	
	VGA colour monitor	
	Ethernet interface	
Facility Requirements	Power :	208 or 240 VAC, Single Phace, 70 A, 60 - 50 Hz
	Venting :	Exhaust vent required
	Nitrogen :	1.5 SCFM @ 50psi
	Compressed Air :	4.5 SCFM @ 60psi

Table 6.1. Hardware parameters SLS Sinterstation 2000

The new powder feeding mechanism uses two cartridges flanking the part-build cylinder. It allows for bi-directional powder feeding to the roller and lays powder across the part-build cylinder and resulting in faster overall build times. This new system design also provides an integrated powder handling capability, by which unsintered powder can be returned to the powder feeding cartridges for recycling and thus extend the use of materials in the SLS process.

Another hardware improvement is the integration of the atmoshperic control unit (ACU) within the system cabinet as a modular component. This unit provides an inert nitrogen flow to the Sinterstation 2000 process chamber, to ensure safe material sintering operations. The earlier SLS Model 125 required additional peripheral equipment to provide the inert capability.

The only requirement for emissions from the system is a single external vent for exhausting very low levels of nitrogen.

As discussed under software improvements, a closed-loop PLC provides the Sinterstation 2000 System with advanced operational control that allows it to operate in an unattended mode, an important capability for users building large, complex part geometries.

The Sinterstation 2000 employs an X Window, Motif-based graphical user interface (GUI) for advanced person-machine interface. This allows user-friendly control of process parameters such as laser energy output, process chamber temperature, and scanning speed. In addition, specific material processing parameters acquired through computer simulations are incorporated in the Sinterstation 2000 system. The parameters are now accessed through the system user-interface and are integrated with the system's programmable Logic Controller (LPC) in a closed-loop configuration that allows the system to operate in a self-correcting, unattended mode.

Other software enhancements include beam compensation, or software-based laser control that results in greater dimensional accuracy in the sintering of part geometries. Also, part previewing and build status are displayed through the system's user interface graphics. These features allow users to view and orient multiple parts within the part-building cylinder before beginning the build process, and to view a graphics representation of the build status of the parts undergoing the sintering process. Finally, the 2000 System's operating software is UNIX-based, which is gaining widespread acceptance for industrial engineering multi-tasking applications.

6.1.3 The process sequence and parameters

Design models created in CAD, CAE or CAM software are entered into the SLS through the STL Interface. The STL file is transferred to the SLS machine. Geometric processing proceeds by first slicing the geometric description of the part into layers representing the sintering planes, typically at 0.076 - 0.508mm intervals. The slicing operation generates the curves or contours that represent the boundaries of the part for each layer.

This cross-sectional information is analysed and converted into information used by the controllers for the scanning mirrors. A very thin layer of heat-fusible powder is deposited into the part build cylinder and heated to just below its melting point. An initial cross-section of the object under fabrication is traced on the layer of powder by the laser (Fig. 6.2). The speed, power, and overlap of the laser beam are all determined by the type of powdered material being used in the system. The temperature of the powder impacted by the laser beam is raised to the point of "sintering", forming a solid mass. The next slice of the computer model is taken by the computer and analysed, while the part build cylinder is lowered by a predetermined amount ranging from 0.076mm to 0.508mm. Another thin layer of powder is then deposited onto the workspace on top of the previously fused layer. Again the laser traces the cross section of the next layer.

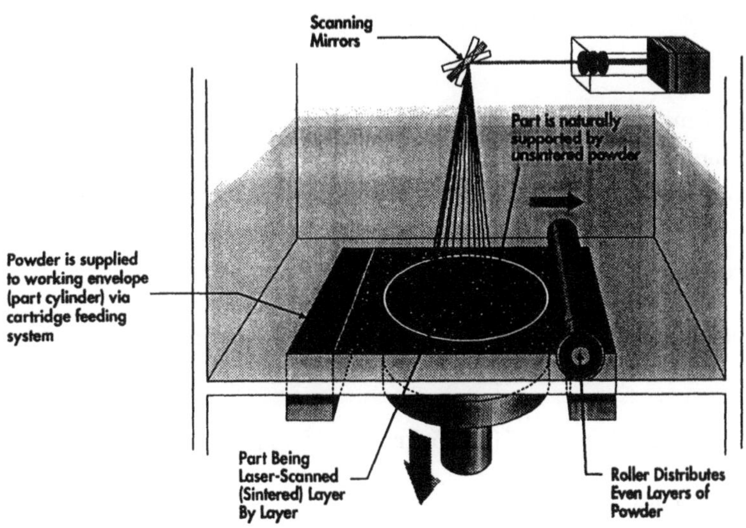

Fig. 6.2. Selectiv Laser Sintering Process

As the process is repeated, each sintered layer fuses to the underlying layer, and successive layers of powder are deposited and sintered until the object is complete.

The essential process parameters which have to be determined are

- the speed, and
- the laser power

Speed

The main factor affecting the speed at which a part can be built is the thickness of the layers being sintered. A part which is being sintered with a layer thickness of 0.125mm (0.005inch) will take approximately twice as long as a part being sintered with a layer thickness of 0.25mm (0.010inch). This is due to the total number of layers needed to manufacture the part. At 0.125mm (0.005inch) per layer it will take 200 layers to build 25.4mm (1inch) of part height. Using a 0.25mm (0.010inch) layer thickness, the 25.4mm (1inch) part will require 100 layers to build the part.

The time needed to scan one layer of a part is determined by the area of the layer, the complexity of the layer, and the number of parts being built simultaneously. As the size, complexity, and number of parts increase, the time needed to complete a layer in the process increases.

The type of material will also greatly affect the time needed to complete a part. Wax requires more time to process, since a cool-down time is added to each layer at the end of the scanning process. This cooling period may be from 1 to 30 seconds, depending on the geometry and size of the parts being formed. Typical build times for materials used in the SLS process are:

- Wax build rate : 3.125 to 12.5mm (0.125 to 0.5inch) of vertical height per hour
- Thermoplastic build rate of : 5.0 to 25.4mm (0.2 to 1 inch) of vertical height per hour.

To increase the overall speed the SLS process software performs concurrent slicing of the part geometry files while processing of the object is taking place. Moreover, no post-production curing is necessary with the SLS process when using wax or thermoplastic.

An additive layer process.

Based on an additive-layer process, SLS technology does not require traditional dies or part-specific tooling and is particularly applicable where fewer, more complex designs are to be prototyped or modelled.

The additive-layer nature of the SLS process also allows for the creation of very complex parts without the need for external support structures, clamping, or repositioning of specific portions of the design geometry. This is possible because the unsintered powder that surrounds the fabricated part during the SLS process provides a customized support structure that is easily removed upon part completion. Designs with internal cavities, overhangs, notches within notches, and other intricate geometries, which are difficult to achieve through other processes, are easily produced using the SLS process. In short, the SLS process allow a degree of freedom of overall design freedom never before realized because of the constraints which tooling problems imposed on designs. The laser beam intensity is modulated to sinter the powder only in areas defined by the object's design geometry. In areas not sintered, powder remains loose and serves as a natural support for the next layer of powder and the object under fabrication. When the part is complete, it is revealed by brushing off the loose powder surrounding the part (or air blowing, see Figure 6.3)

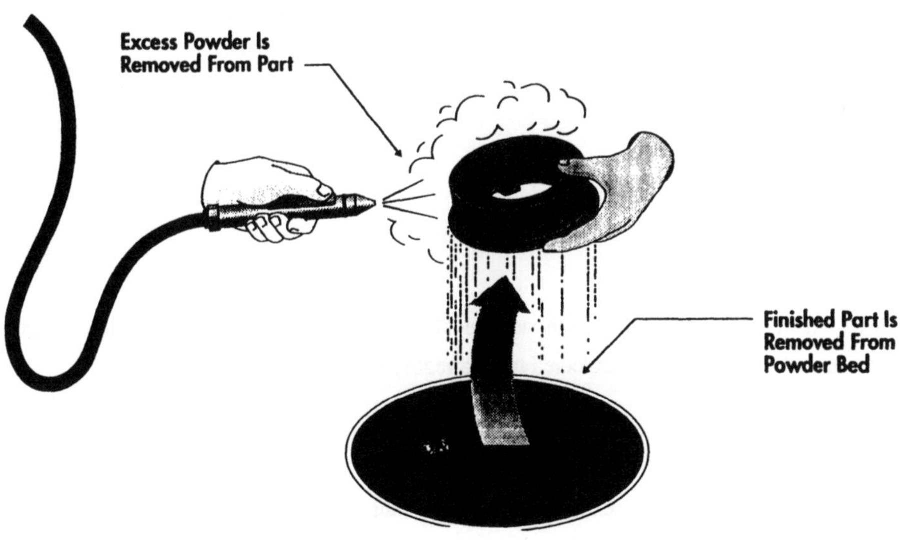

Fig. 6.3 Cleaning of SLS parts by air blowing

6.1.4 Technological and economic features

The essential technological feature of the SLS procedure is the possibility for using multiple materials.

- Virtually any material that softens and has decreased viscosity upon heating can, potentially, be used.
  - polycarbonate
  - investment casting wax
  - PVC
  - ABS-plastic

    } These materials have been successfully used in the SLS process
  - Nylon
- Additionally, ceramic and metal parts can be fabricated with the SLS process by polymer-coating these materials, with curing as a secondary process

The natural evolution of the SLS technology is also heading towards the use of directly sintered powdered metals and alloys. The first successfully results were obtained at the University of Texas at Austin using the following sintering materials:

- brass
- copper
- aluminium
- cobalt

These efforts are continuing with additional metal alloys and metal powder blends. Commercial use of metals is expected to be a reality in the near future.

The direct method of Selective Laser Sintering involves using a higher temperature process chamber and a higher wattage laser to fuse metal and ceramic particles. This process is being used on an experimental basis at the University of Texas in Austin. The University, through DTM funding, has constructed a High Temperature Workstation (HTW) which employs a 1000 watt laser and a process chamber capable of reaching 1000 °C (1832°F).

Using the direct method of sintering, no post-sintering is required to complete the parts. The goals of this research programes are the determination of the optimum process parameters for materials used in the next generation of SLS systems, which will sinter metals and ceramics for future applications that include the direct manufacture of prototype tooling and complex parts.

The areas of current research with metal and ceramics are summarized in the following table:

	Sintering Method	
Material	Indirect	Direct
Metal	Polymer Coated Copper	Cobalt-Tungsten Carbide Copper-Nickel Aluminium-Silicon Carbide
Ceramic	Polymer Coated: Alumina Glass Zircon Alumina/Ammonium Phosphate Alumina/Boron Oxide	Aluminium Aluminium-Oxide

Table 6.2 Future extensions of the SLS procedure

Another important technological feature is the user support by specific software. DTM has improved the basic SLS process through the implementation of a statistical process control program (SPC). This ongoing SPC program has allowed DTM's Development Engineers to understand better the basic process parameters such as material particle size and distribution, thermal dynamics scanning speed, and geometry-specific variables, all of which vary with each material used in the SLS process. This SPC program includes the analysis of material coupons and tensile bars.

6.1.5 Application range

Based on the expanding and diverse variety of powdered materials, there is a relatively broad application range in three major areas /22/

- visual representations (models)
- functional, durable prototypes; and
- production capabilities, including the creation of the moulds, patterns, and tools required for limited production runs and on-demand batch manufacturing

Typical production applications are the following.

Short Run Tooling and Soft Tooling.

Here, the SLS process can be used to create a master pattern from which various types of soft tools, including those made of silicon rubber and epoxy, can be made. These tools can then be used to generate short runs of parts for use in testing, marketing, and verification of fit and form. These short run parts can be used, in some cases, to meet custom manufacturing requirements.

Spray Metal Tooling

An SLS-created pattern made of polycarbonate is coated with a thin layer of protective material, followed by sprayed-on layers of liquid metal to further coat all surfaces. The metal-coated pattern is then put in a container filled with molten metal. The metal cools, and a tool for casting additional parts is created. Because the tool is made of metal, usually zinc with epoxy or aluminium backing, it is more durable than soft tools and can produce runs of up to several thousand parts, depending on the application.

Plaster Cast Tooling

As with other tooling methods, an SLS-generated pattern can be used to create a plaster cast tool, which is then used to make steel and other metal parts.

Vacuum Form Tooling

The SLS process can be used to create a master form, from which the vacuum-formed object will achieve its final shape. During the creation of the form in the SLS process, the form's porosity can be adjusted.

This is an inherent advantage, because a master form created with a suitable level of porosity eliminates the extra step of drilling the dozens of tiny holes needed to allow air to pass directly through the porous material.

Numerically controlled (NC) tooling and path verification

An SLS-generated part can be used to trace and program the tooling paths that operate numerically controlled machines.

Electro Discharge Machining (EDM)

Upon availability of carbon materials for use with the SLS process, it will be possible, quickly and affordably, to generate the carbon or graphite electrodes used in electrode EDM. Currently, complex designs of these electrodes are difficult and time-consuming to produce.

Forging Dies

Drop-forge dies are made of steel or steel castings at a grade of 0.60 percent carbon open hearth. Dies of this steel will forge mild steel, copper, and tool steel. SLS-generated wax patterns can be used to produce forging dies of steel castings. A pair of these dies can then be used to produce up to 20,000 pieces. An overview of the different application ranges is shown in Fig. 6.4.

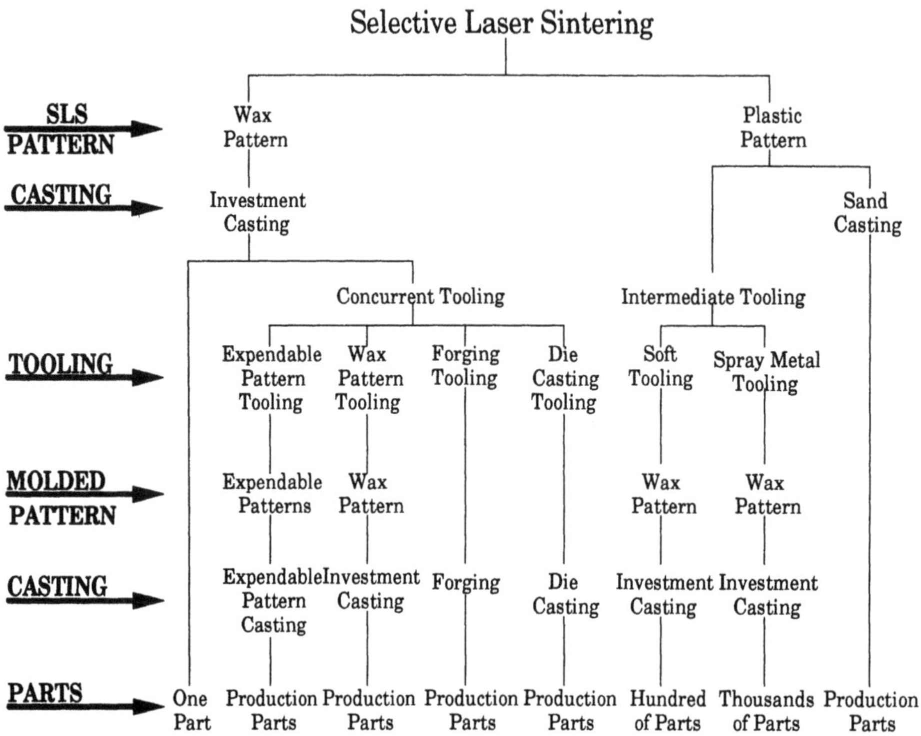

Fig. 6.4. SLS & Metal Manufacturing

The SLS process and wax pattern-making are extremely well-matched, for a number of reasons. These include:

1. The patterns made from the investment casting wax in the SLS process can go directly into the investment casting process, and can follow the procedures normally used for patterns generated via moulding wax. Thus the problems of long burn-out cycles associated with photopolymer resins are eliminated.

2. Accuracy within the specifications of most investment casting applications. The average tolerance range of wax patterns made in the SLS process range from ±0.125 to ± 0.25mm (± 0.005 to ± 0.010inches). This meets the accuracy requirements for most investment casting applications.

3. A surface finish that is suitable for most investment casting applications, average typically 100 to 120 microinches RMS. When smoother finshes are required, secondary polishing and finishing may be employed.

4. Eliminating the tooling stage translates into numerous benefits for investment casters, including minimized upfront tooling costs and significant time savings for prototypes, plus the ability to produce complex shapes that were previously too difficult to produce via simple tooling.

6.1.6 Advantages and disadvantages

The most important advantage of the SLS procedure is the variety of materials usable, both currently and in the future.

- Numerous materials can be used, including most engineering plastics and standard investment casting waxes. This relates well offers to other traditional manufacturing processes

- Production, from powder to part, is generally a same-day process. Parts are produced at a rate of up to one inch per hour.

- Complex designs unattainable by other methods can be successfully created, quickly and economically.

- Parts are of full strength and do not reqire additional support during the building process or subsequent postcuring

The SLS process offers great strength and toughness. Important disadvantages result from the specific procedures:

- during the solidification it may happen that the borderline additional powder is hardening. That is an important influence on the true accuracy

- in general, the surface smoothness is restricted by the size of the powder particles and the laser spot. Surface finishes of 20-60 microns RMS can be achieved only by post-finishing the parts.

- the necessity to provide the process chamber continuously with nitrogen to ensure safe material sintering operations leads to additional expenditure

- depending on the size of the powder particles and the above-described hardening effects, the surface is not always smooth and level

- toxic gases emitted from the fusing process have to be handled carefully (this is specially the case concerning plastics like PVC)

6.2 Fused Deposition Modelling (FDM)

6.2.1 Functional principle

The FDM process uses the Stratasys so called 3-D Modeller in conjunction with a CAD workstation. The 3-D Modeller operates on the principle of a three axis NC-machine tool. A nozzle - Stratasys uses the term FDM-head - is controlled by a computer along three axes. The nozzle guides the specific material (plastic filament), which is melted by heating. So the material leaves the nozzle in a liquid from, which hardens immediately at the temperature of the environment. This builds the model upwards from a fixtureless base.

6.2.2 Hardware

The hardware with computer and 3-D Modeller is suitable for an office environment, measuring 762 x 914 x 1727 mm (30 x 36 x 68 inch). The main components are demonstrated in Fig. 6.5. The remarkable speed of the FDM head; of up to 22860 mm/min (900 inches/min), is based on its light construction.

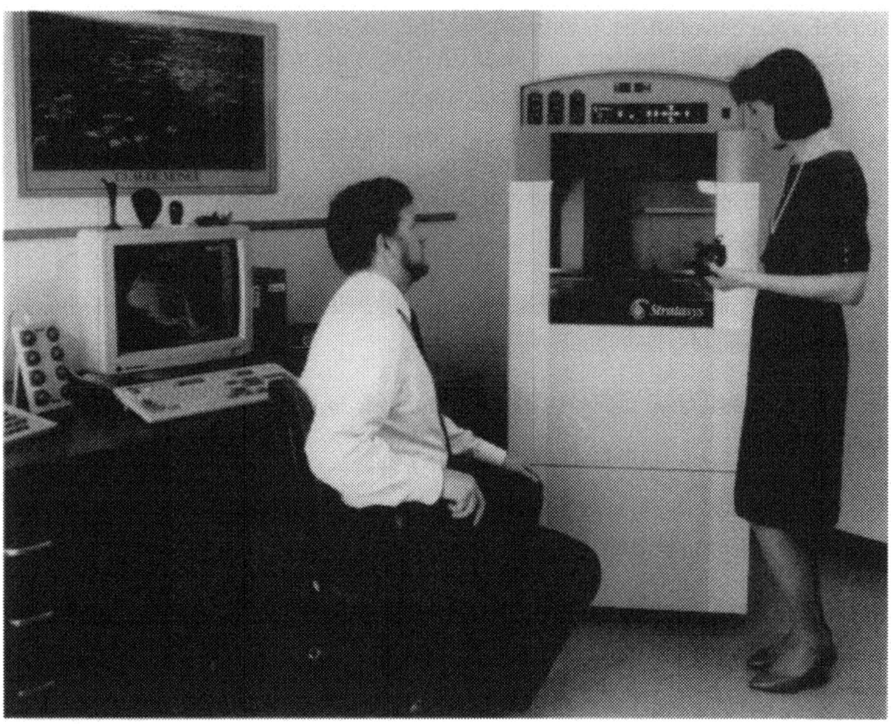

Fig. 6.5. 3-D Modeler

6.2.3 Process sequence /23/

The general process sequence is similar to all other SFM technologies. Input data for the 3-D-Modeller could be:

- IGES format (this file can be edited, scaled, oriented, and even surfaced if required. Supports, if needed, can be added graphically)

- direct, through NC code (taken directly from other CAM programs)

- through the Stratasys' Proto Slice software, the stratasys system uses a special packaging of CAD/CAM software.

  Additionally a full CAD package is provided as part of a total turnkey system, allowing the user to design and build a part completely from scratch. The capabilities include a full NURBS-based surface modeller for complex or simple shape creation.

- through STL format, which can be transformed into NC code

Liquid thermoplastic material is extruded and then deposited in ultra-thin layers from the lightweight FDM head, one layer at a time. This builds the model upward from a fixture-less base. The plastic or wax material then solidifies in O.1 second as it is directed into place with an X-Y controlled extrusion head orifice that creates a precision laminate.

A spool of modelling filament 1,27mm diameter feeds the FDM head, and can be changed to a different material in 1 minute.

Maintaining the liquid modelling material at a temperature just above the solidification point is fundamental to the FDM process.

The thermoplastic melt temperature is controlled to 1 degree Fahrenheit above solidification.

The 3-D Modeller tightly controls the thermoplastic melt temperature just above solidification, then employs precision volumetric pumping through the extrusion orifice.

The model is immediately sheared, quickly solified, and each layer bonded to the previous layer through thermal heating. The completed model is ready for removal from the 3-D Modeller(see Fig. 6.6).

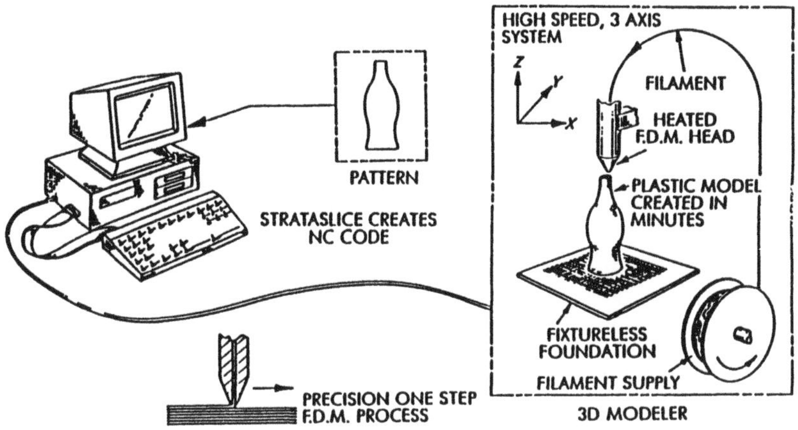

Fig. 6.6. Fused Deposition Modeling - Stratasys

Successive laminations, within the 0,0254mm to 1,27mm thickness range with a wall thickness of 0,254 to 6,35mm, adhere to one another through thermal fusion to form the model. The overall tolerance is ±0,127mm in the X,Y,Z axes over a 305mm cube working envelope.

6.2.4 Technological and economic features

The important technological features are the relatively high speed operation and the safety benefits of the materials used. The system uses a wax-filled plastic adhesive material, a tough nylon-like material or investment wax. These thermoplastics soften and liquefy when heat is applied.

The Stratasys investment casting is an industry standard casting material. This wax rapidly detaches from the shell, using normal investment casting procedures, and provides superior surface finishes. Because this is a standard investment casting wax, it can become part of the foundry's normal recycling process. The machinable wax, a wax-filled plastic adhesive material, is primarily used for conceptual modelling and spray metal moulding. For instance, the accuracy of the model allows its use in the spray metal process for injection prototyping. Both the investment casting wax and the machinable wax can streamline the manufacturing process by allowing the user to go directly to soft tooling, using the model as the pattern for the investment casting or the spray metal injection moulding processes. The plastic filament is a tougher material, producing a sturdier model suitable as a concept model or for form, fit and some functional applications. It is strong yet flexible, with good surface finish. The materials to produce a part are cost effective, usually under twenty dollars per part.

Repeatability measures how closely the machine can run the same tool part 100 times, with the part remaining within tolerance and resolution. The Stratasys Inc. system uses a spool-based filament system to feed the material into the modeller. The spool based system has proved to be a significant strength of the 3-D Modeller. The spools give the user the ability to change material in about one minute, by threading the desired material into the prototyping unit. The previous material is purged out of the orifice in about 5 minutes. There is virtually no waste and no vat to clean out; only what is required for the model is used.

Based on the high speed operation, it is possible to produce small models in times of 30-60min.

The materials to produce a part are cost effective, usually under ten dollars /22/. So the FDM procedure seems to be one of those currently cheapest.

6.2.5 Application ranges

The application ranges lie in

- conceptual modelling
- fit, form, and functional applications and models for further manufacturing procedures
- investment casting and injection moulding subject to the above characteristic overall tolerance

Some examples are shown in figure 6.7.

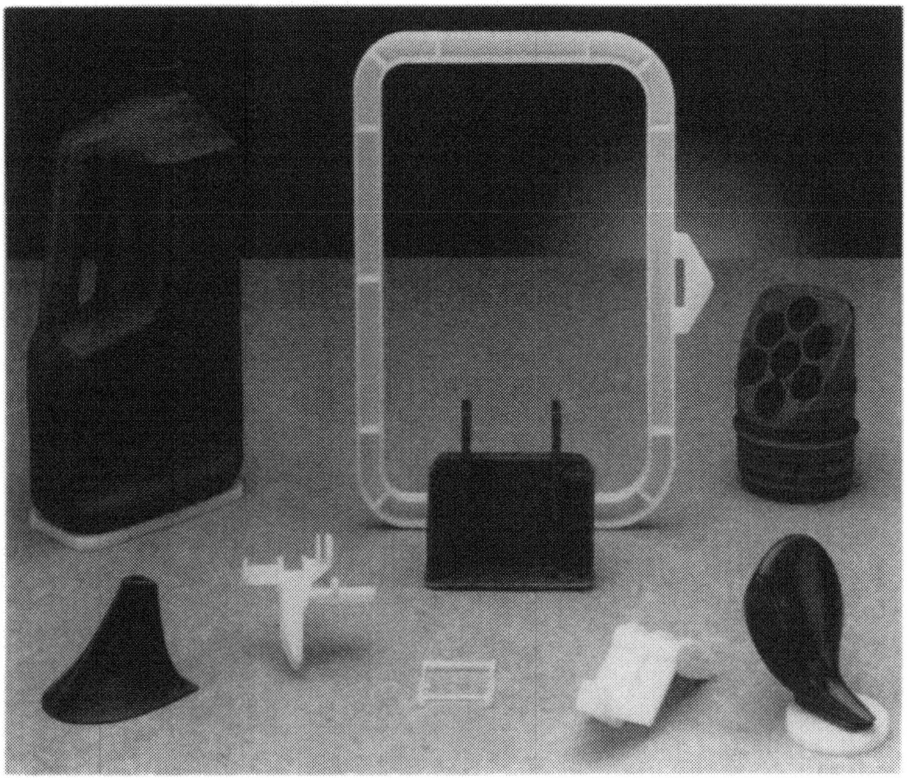

Fig. 6.7. Examples FDM-process

6.2.6 Advantages and disadvantages

The most important advantages are:

- the quick and cheap generation of models
- the easy and convenient data handling
- the safety benefits. There is no worry of possible exposure to toxic chemicals, lasers, or a liquid polymer bath
- the procedure requires no cleaning and produces no waste, and no post-curing is required
- The simplicity of the FDM technique allows for quick and easy installation

The disadvantages are due to restricted accuracy. A physical restriction is given by the use of wire material of a diameter 1,27 mm, which is far more than the laser beam spot size. No information about shrinkage and distortion behaviour is available. Based on some demonstrated examples, it can be concluded that, mainly, compact parts can be manufactured.

6.3 Laminated Object Manufacturing - LOM

6.3.1 Functional principle

For manufacturing of 3-D objects based on 3-D geometrical data, the LOM principle is relatively easily understood. The starting-point is sliced data of the object, which is used for controlling a laser beam which cuts the contours of foil materials. During the process these foils will be glued together and the desired model is created layer by layer.

6.3.2 Hardware /24/

Figure 6.8 demonstrates the fundamental structure of a current LOM machine.

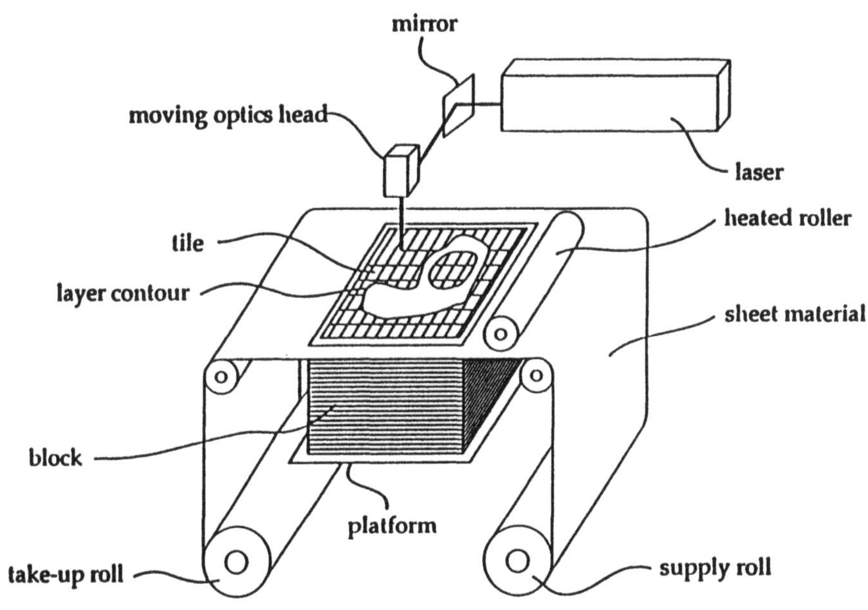

Fig. 6.8. Laminated Object Manufacturing

The mechanical part of the system contains an unwinding and a rewinding roll connected by a ribbon of sheet material, routed via several idler rollers. These rolls store and supply the material. The laminated part is grown on a platform capable of a vertical incremental movement under the action of a stepping motor. Above the platform there is a heated roller, capable of heating and compressing the ribbon between itself and the stack of laminations on the platform. As a result of a single reciprocal motion of the heated roller the ribbon material is bonded to the top of the stack. An X-Y positioning table carries two mirrors which reflect a beam from a CO_2 laser and a lens which focuses the beam on the upper surface of the laminated stack, so as to cut the very top layer. Scrap pieces remain on the platform as the part is being built. They are diced by the laser beam into cross hatched squares and serve as a support structure for the part.

6.3.3 Process sequence

The working of an LOM machine is as follows.

- The computer which runs the system is capable of slicing a 3-D solid model into thin two-dimensional cross-sections. The thickness of each cross-section is equal to the thickness of the material used in the process.
- The geometrical information is fed into the LOM system, which guides the laser beam cutting around the periphery of the cross-section. The laser cuts to the depth of one layer of the material. The material which surrounds the cross-section is usually cross hatched by the laser beam into squares, to facilitate separation at the completion of the process (Fig. 6.9).

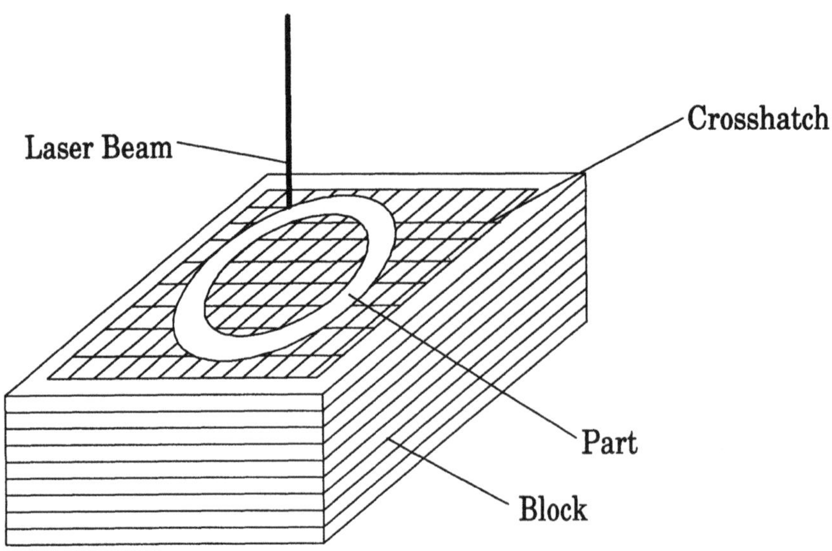

Fig. 6.9. Crosshatch

In the areas where attachiment of the material comprising the part to the material which surrounds it is undesirable, the laser beam ablates the portions of the bottom layer to prevent bonding. Sometimes a fine cross hatch of these portions can achieve virtually the same effect.

- Next, the platform moves down. The ribbon moves, by an increment exceeding the length of a cross-section, onto the rewinding roll. As the platform moves up, the heated roller moves across the stack while pressing the ribbon against the stack and bonding it to the upper layer.
- Meanwhile, the height of the stack is measured by the machine and fed back to the computer, where the next layer is calculated according to the current measurement.
- The laser beam cuts a new cross-section. The process continues until all of the cross-sections have been deposited and cut. This cycle is shown in Fig. 6.10.

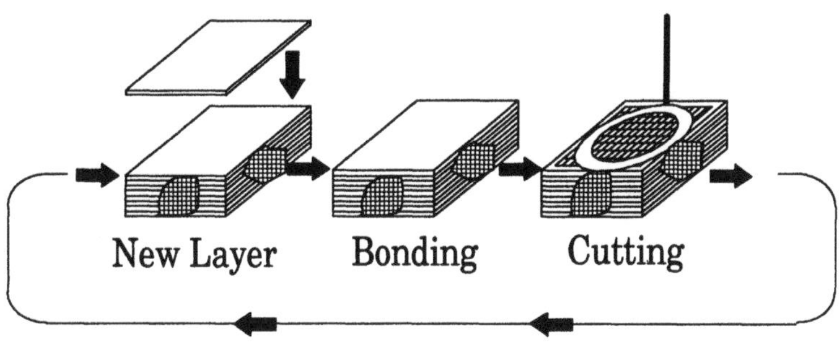

Fig. 6.10 LOM Cycle

- The product comes out of the machine as a rectangular block containing the part (Fig. 6.10). The material surrounding the required part has already been diced into small cubes due to a "cross hatch" cut by the laser, so they can be separated easily from the part.

6.3.4 Technological and economic features

An important feature is the possible application of different materials, like

- paper
- plastic
- ceramic
- composite

Virtually any sheet-form material which can be bonded to itself and can be cut by a laser is suitable for this technique.

Because of the simple concept, the price of LOM equipment is relatively low and also the overall process costs are below many comparable costs.

6.3.5 Application range

Based on the specific aspects of the procedure, the main application range is the working area of conceptual design.

6.3.6 Advantages and disadvantages

Specific advantages emphasized /24/ are that a variety of organic and inorganic materials can be used in the LOM process allowing the user to meet various requirements such as price range, environmental concerns, chemical and mechanical properties, colour and appearance, secondary process durability, etc.

The process is much faster than competitive techniques, since the laser does not have to scan the entire area of a cross-section but just go around its periphery.

Layers are glued to the stack virtually instantaneously, substantially reducing layer formation time. The speed advantage of the process grows with the size of the manufactured part. The more volume of material contained within the part, the greater is the speed advantage.

The process produces virtually no internal stress and associated undesirable deformation. Also, because the in-process part is already solid, it is convenient to measure the actual height and perform real-time slicing on the fly. High precision (below 0.127mm _ .005") is achieved due to the absence of shrinkage and ability to perform real-time measurement and slicing.

The important disadvantages are:

- the stability of the objects seems not sufficient to allow applications other than design support

- because the procedure is not well suited for manufacturing parts with thin walls in the z-direction, the application range is more or less restricted to complex parts.

LOMSlice possesses the robust capacity of dealing with imperfect STL files. Not every STL-generating CAD system does a perfect job; some STL files are created with discontinuities, which means that the surfaces of STL objects are not closed completely. LOMSlice can tolerate these cases and clarify these ambiguities to make a successful slice.

6.4 3-D-Printing /25, 26/

6.4.1 Functional principle

Three-dimensional Printing creates parts by a layered printing process, based on sliced 2-D object data. A specific two-dimensional layer is created by adding a layer of powder to the top of a piston and cylinder, containing a powder bed and the part being fabricated. The new powder layer is selectively joined where the part is to be formed by "ink-jet" printing of a binder material. The piston, powder bed and part are lowered and a new layer of powder is spread out and selectively joined. The layering process is repeated until the part is completely printed. Following heat treatment, the unbound powder is removed, leaving the fabricated part.

6.4.2 Hardware

The experimental system available at MIT consists of an 80386-based computer and control software, a DCJ 4000 motion controller, a DC stepping motor driving an x-y table, a cylinder regulating the depth (z-axis) of the powder bed, and a binder dispensing system based on continuous-jet ink-jet printing. The printing determined speed by the raster scanning along the y-axis is fast (1.5m/sec) and blow along the x-axis (see Fig. 6.11)

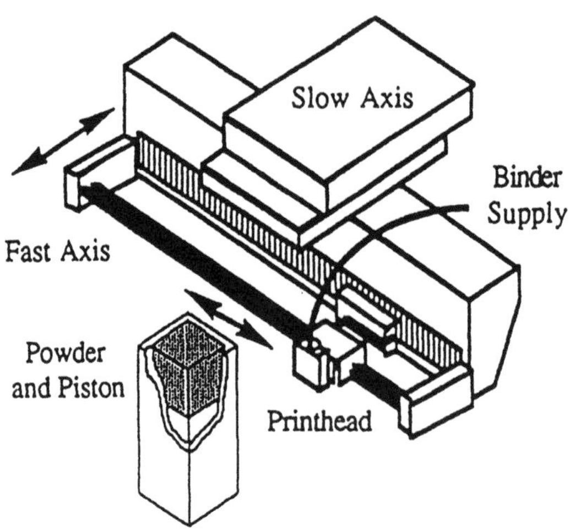

Fig. 6.11. The 3D Printing System

The continuous-jet nozzle has a circular opening of 46μm diameter. The binder stream is broken into droplets 80μm in diameter by a vibrating (60 kHz) piezoelectric transducer.

The printing pattern from the CAD data is mapped into the amount of negative charge carried by each droplet; a fixed electrostatic field steers these droplets to the area of powder to be bound, at a vertical speed of 10 m/sec.

6.4.3 Process sequence

The entire process chain starts like other SFM procedures also using 3-D object data.

Individual layers are created by spreading a layer of powder, and then selectively joining the powder in the layer by ink-jet printing using a binder material.

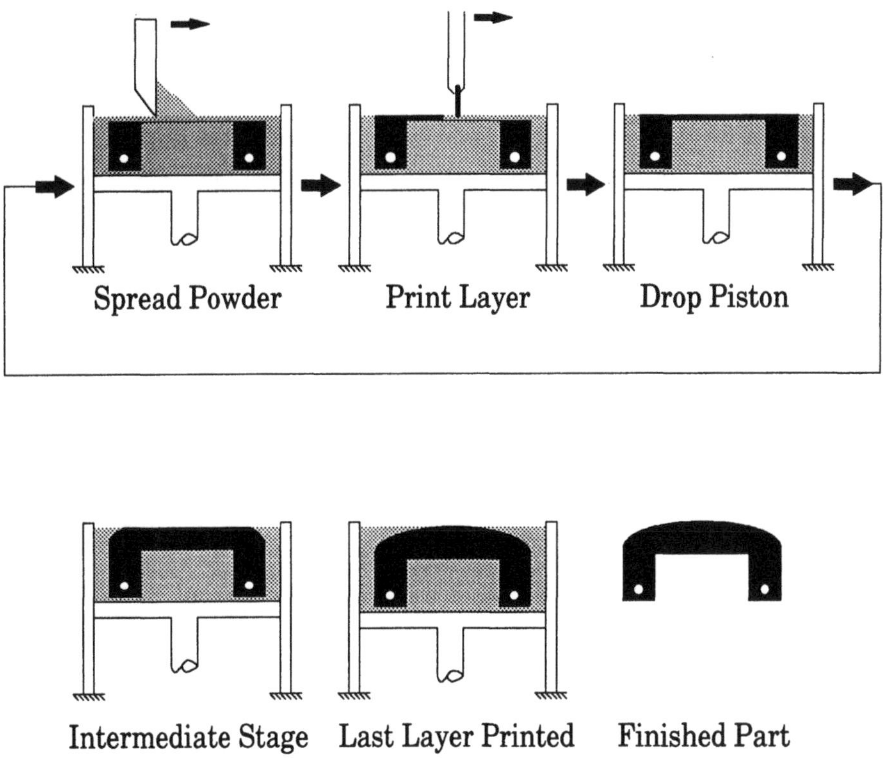

Fig. 6.12 Sequence of operations in the Three Dimensional Printing Process

The binder gives the printed component enough strength to withstand the handling involved later in the process. After being printed, the component will be either fired or sintered, depending on the type of working material, to reach its maximum strength.

6.4.4 Technological features /26/

In principle 3-D Printing can form any material that can be obtained as a powder which could mean almost any material. Because different materials can be dispensed by different printheads, 3-D printing can exercise control over local material composition. The proper placement of droplets can be used to create surfaces of controlled texture, and to control the internal microstructure of the printed part. Ceramic moulds for metal casting and tooling for forming processes such as injection moulding are the two application areas first used for 3-D Printing. When applied to make ceramic moulds, colloidal silica is the binder material used to join alumina powder. The result is the direct production of a casting shell with integral parts which can be used to create a hollow casting in almost any geometry. Since the materials used for the 2-D Printed shell are essentially the same as those used to make shells by the lost-wax process, a wide variety of material may be cast. Metal tool such as those used for injection moulding, thermoforming and other processes can be printed by spreading a metal powder and joining it with an organic binder which is later hardened by burn off. The roughness values attained are within the range 1.3 to 3.0mm. The normal applied layer thickness is in the range 76µm to 254µm.

6.4.5 Application range

Because of the current experimental stage reached no detailed explanation are possible.

The initial applications of 3-D Printing are:

- The direct fabrication of ceramic cores and shells for metal casting.
- The direct fabrication of porous ceramic preforms which when infiltrated by liquid metal will form metal-ceramic composite parts.

6.4.6 Advantages and disadvantages

Evaluation of these aspects is only realistically possible for commercial and industrial applied systems. But the almost unrestricted application of any kind of material seems an important advantage.

A disadvantage results from the granular nature of the material and the interactions between the binder and the powder. This leads to obvious textures of all surfaces.

6.5 Ballistic Particle Manufacturing - BPM

Because this principle is still under development, only a short description is possible. The process uses three-dimensional solid model data to direct streams of material (photocurable polymers, waxes, plastics, ceramics, or metals) at a target. The three-dimensional objects are generated in a manner similar to that of an ink-jet printer producing two-dimensional images. BPM builder an object by a 3-axis controlled NC-system, which "shoots" the material onto a target by a piezoelectric ink jet mechanism. A continuous bead of UV curable polymer is extruded onto a base or onto previous areas of hardened material, and cured by continuous ultraviolet flooding.

7. General Tasks for Process Realization

7.1 The process chain for SFM procedures and integration aspects

It should be emphasized that the general process sequence is similar for all kinds of SFM procedures, as is shown in Fig 7.1.

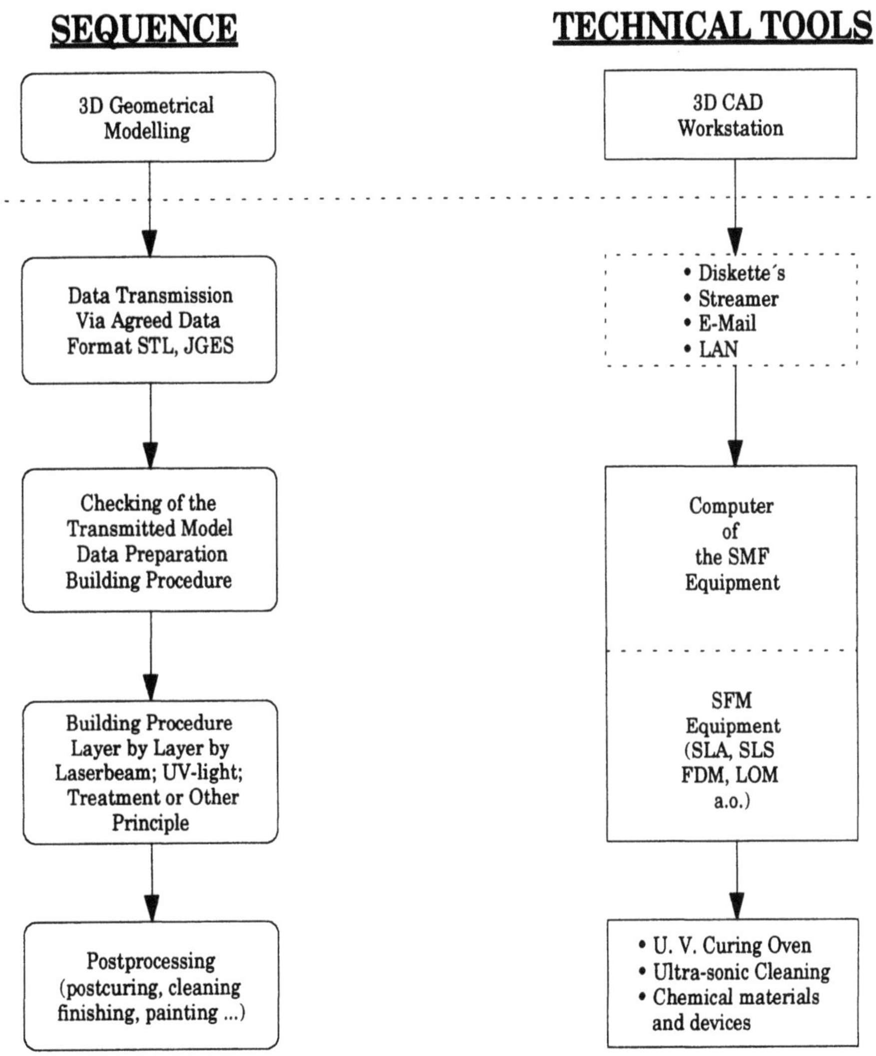

Fig. 7.1. General Process-Chain and Technical Tools for SFM Process

This allows interested users to employ the same kind of data preparation and, in the case of decision-making, the most suitable process for the testing of different procedures (see Chapter 11).

Note that the SFM systems are closely related with all other relevant tasks in a Computer Integrated Manufacturing environment, which means that the available CAD/CAM systems have to be extended to allow the essential function interfacing to SFM techniques (see Fig. 7.2).

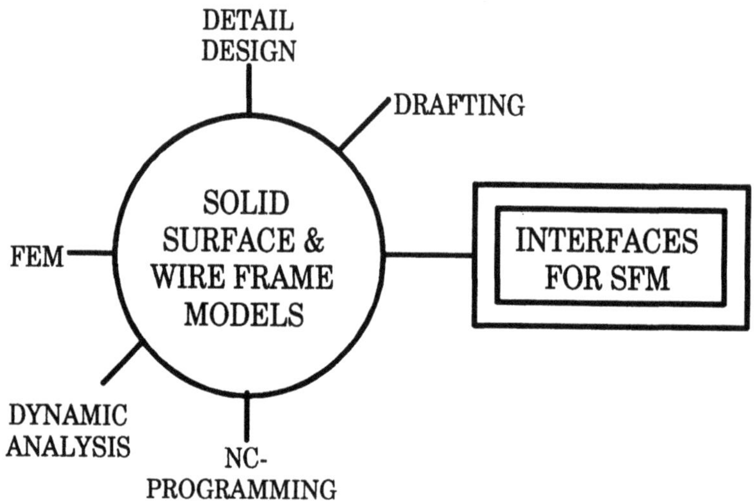

Fig. 7.2. CAD/CAM Basic Systems and Extensions for SFM

The involvement of powerful 3-D CAD/CAM systems allows the close links to all other design and manufacturing tasks.

7.2 Geometrical Modelling

Advanced 3-D CAD Modelling is a general prerequisite and, in connection with SFM procedures, the most time-consuming part of the entire process chain.

It is most important that such 3-D geometric models can be shared by the entire design team for many different uses, such as interference studies, stress analyses, FEM analysis, detail design and drafting, planning, of manufacturing, including NC programming, etc.

Most CAD/CAM systems now have a 3-D geometrical modeller facility with the above-mentioned performance.

For a detailed explanation of the suitable CAD/CAM systems that are available, and specific instructions and handbooks from particular vendors, see /49/.

From the aspect of user-friendly and efficient utilization of SFM systems, some essential requirements must be considered.

7.2.1 Basic terms and methodologies

For understanding 3-D geometrical modelling, three basic terms are essential /49/.

- object modelling
- construction geometry
- feature definition

The geometry of an object can be created in several different ways. These include primitives (geometry from familiar solid shapes); extruding or revolving a 2-D profile, and skinning (using a set of profile cross-sections to define a solid).

After solid models of objects are created, they can be modified by construction operations which include cutting, joining, and intersecting with different objects. A die cavity can be created, for example, by cutting the part geometry from the die blank. Object geometry can also be modified by changing its dimensions or by changing the primitives or profiles which were used to create the object.

The Construction Geometry tasks allow an engineer to begin creating geometry with 2-D sketching or drawings and then build 3-D solids from this geometry. It is possible to create three kinds of construction geometry: wireframe, profiles, and patches (skin groups).

The following diagram illustrates the different ways that the three different types of construction geometry can be used to create objects.

In most cases, solid objects will be created to represent the final design, so construction geometry can be considered as a kit of different tools for creating solid objects.

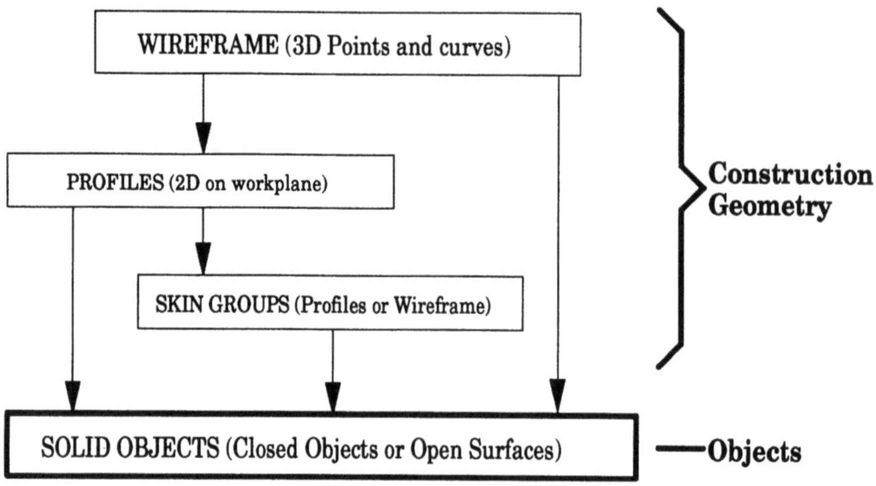

Fig.7.3. Creating Solid Objects From Construction Geometry

A wireframe is a collection of points and curves (lines, arcs, circles and splines) in three-dimensional space. Wireframe geometry may be used at the beginning of a design to allot space for different parts in an assembly. This geometry can then be used as an aid in creating the part geometry to fit where it belongs in a 3-D layout. Wireframe geometry can also be used to create profiles and skin groups for more complicated solid geometry.

Profiles are two-dimensional points and curves connected end-to-end to define 2-D outlines. Profiles can be created directly on a workplane oriented in three-dimensional space, or points can be projected from a 3-D wireframe or solid object geometry. Profiles can be controlled by variational geometry constraints. This allows a designer to sketch a profile freehand and later create or modify dimensional or angular constraints to create the desired shape. Profiles can be used to create solid objects by extruding or revolving one profile, or skinning between multiple profiles.

Skin groups are collections of profiles representing cross -sections through a part, or collections of wireframe curves defining a surface.

Features-based modelling makes it easier to create solid models and easier to make changes after the model is built. Features are predefined objects or construction tools which prompt users for key parameters. The concept of features makes the modelling process match the physical reality of the manufacturing processes more closely,rather than matching the mathematics of a solid modelling program. For example, a counterbored hole can be constructed with a feature called a "Counterbore" which corresponds in size and shape to a milling tool instead of a pair of mathematically-abstract cylinders.

Features can be used for simple construction tasks such as making slots and ribs, or to create an entire featured object such as a bracket.

The next step after solving the modelling task is the storage in a computer. This task can be considered and solved under three headings:

- solids

- surfaces

- wireframes

A solid object must contain enough information to describe fully the boundaries, surfaces, and the topology of the inside and outside of the object.

A surface model stores equations representing each of the surfaces of an object, as well as the edge boundaries. The fancy animated 3-D graphics generated for television "flying logos" use only surface modelling, since mass properties or other mechanical applications are not a concern. In a surface model, it may still be difficult to check for interferences, since we may not know for sure which is the inside or the outside of each surface. For the same reason, it will be difficult to compute mass inertial properties of such an object, especially if the surface model is not completely closed /see Section 7.2.2).

In a wireframe model , the equations of the boundary curves of the object are stored. With just a wireframe representation, it is impossible to check for three-dimensional interferences, since there is no representation for the surface between wires.

Topological and geometric data /50/

The topological and geometric data the are two types of data used in describing a shape geometry. Topological data describe the connecting relationships among geometric elements, such as the vertiees, edges and faces in the boundary representation model. The topological relationships among these elements will remain unchanged through "rubber sheet" transformations. That is the properties of geometric configurations are not changed by deformations such as twisting, scaling, inflating, stretching etc.

The geometric data consist of the basic shape parameters and all information about the geometric shape of an object, including its position in space and the precise geometrical location of all aspects of its entities. Examples of geometric data are the coordinates of vertex points, and position, direction or geometric coefficients of surfaces or planes. The geometric data specify a spatial subset in 3-D space.

Geometry Representation /49/

Most commercial solid modelling programs are based on either Boundary Representation (BREP) or Constructive Solid Geometry (CSG).

BREP modellers store the results of operations (i.e. what exists after creation and manipulations), and CSG modellers store a history of objects and operations required to construct the part.

For any application of solid modelling techniques specific requirements have to be considered. It must be emphasized that for SFM applications, it is necessary to consider some rules and requirements.

7.2.2 Rules and requirements for efficient SFM utilization

7.2.2.1 Basic rules and requirements

Related to the complexity of the parts, it seems useful to consider a specific classification by size. Such a classification can be based on the number of primitives (shape elements) plus features, or the number of Booleans.

In agreement with Lowrey /51/, subdivision in three classes is possible.

Class Booleans description; Examples

I	< 75	some complex surfaces; piston, connecting rod, cam, crankshaft
II	50 - 200	complex surfaces; manifold, windage tray, larger housings
III	> 150	highly complex surfaces and constructions; engine block, cylinder head

The most important necessity for SFM applications is closed geometries. This can be characterized by the term clean modelling /51/.

Clean modelling techniques allow the user to avoid the creation of excess geometry, and always take precedence over all other rules. And, if unclean surfaces are created, techniques exist for removing such geometry.

Clean models have continuous surfaces wherever possible. This requires that the surfaces have the fewest possible vertices, edges, surfaces, gaps, deviations, etc. The constructions must be chosen so that surfaces are continuous. Do not break a continuous surface unless absolutely necessary.

Clean surfaces are also required for finite-element meshing and other applications. If a surface is not clean, the NC tool path generation will require stepping over extraneous edges and surfaces.

Some other important application requirements can be characterized as follows /52/.

Reliability: Although it is difficult specifically to rank the requirements, reliability has to be at the top of the list. There are two obvious failure paths for any automated solids application. Either the modeller fails, or the algorithms in the application fail. Certainly, the best application algorithms in the world are of little use if the underlying modelling technology is not stable.

Precision: As a basic technology for purposeful applications, precise modellers that operate to double precision accuracy are essential. Advances made with new modellers have shown that is possible to meet or exceed the performance of faceted modellers with precise modelling technology.

Parametric Surfaces: Early modellers typically stopped short of providing general parametric surface support. The modelling of a car body panel, plastic injection mould, aerofoil, etc. fall in the category of parts that demand more sophisticated modelling. Without parametric surfaces embedded in the modelling technology, support for general purpose applications cannot be realized.

3-D Entities: Whereas early modellers supported only the modelling of fully bounded solids, advanced applications require support for the modelling of a variety of 3-D entities. This includes bounded and unbounded curves and surfaces as well as complete and incomplete solids.

Performance: Modellers must be able to work at interactive speed in support of advanced applications. Performance improvements must be 5-20 times that of early solid modelling systems to be considered sufficient for advanced applications.

Product Data: Modellers must be capable of handling more than just shape information. Most modellers have supported the attachment of product data in the form of attributes, but the class of attributes has usually been predefined using simple data such as colour, surface finish and textual fields. In addition, modellers have predefined rules for managing attributes. Applications need to define attribute data and the rules for managing attributes.

Kernel Functionality: A Kernel modeller should be viewed as a geometric operating system. That is, a Kernel should be application-independent, and provide a rich set of facilities for modeller extension. In this way, the modeller becomes a facilitator of applications development rather than creating obstacles for the application.

Non-Manifold Objects: It is often the case that no physically realizable solid is required. During the creation of a model, curves are created, closed into faces and perhaps extruded into solid. It may be that one wishes to construct a certain geometry in wireframe or surface form. It may also be that a fully bounded solid is subdivided into a non-manifold object, for unforeseen reasons. Perhaps it is for modelling composite materials, for the creation of celluar objects in support of finite-element modelling, or for ease of creation of user-friendly interfaces. In any case, if the modeller cannot provide a consistent representation for curves, surfaces and solids, it remains a barrier in the way of the application.

Programming Interface: A programming interface to solid modellers is required for any application development. The programming interface must be accessible from a variety of programming languages.

Parametric Curve and Surface System: The Parametric Curve and Surface provides classical curve and surface functions, including intersection, evaluation and analysis. Because it is separated from the Kernel by a procedural interface, other curve and surface systems that adhere to the spline interface can be substituted. Analytic curves and surfaces can be handled by the Kernel to improve modeller performance. An application that accesses the curve and surface subsystem through the API can remain independent of other surface implementations in the modeller.

Spline Interface: The Spline Interface is a set of routines that offer a procedural access to the Parametric Curve and Surface system. The Kernel interfaces to the Parametric Curve and Surface system only through the Spline Interface, and therefore need not know the particulars of the underlying curve and surface system. It knows only how to interface with it. Through this interface, curves and surfaces are encapsulated as types known to the Kernel and API. The integration of an alternative curve and surface subsystem requires that the spline interface be reimplemented.

Kernel: The Kernel Modeller is the heart of the geometric modelling system. It is a boundary evaluator that primarily handles Boolean operations, modeller data structures, memory management, and event logging (bulletin board). The Kernel is responsible for modelling analytic curves and surfaces, to improve modeller performance. In this way it need only interface with the Parametric Curve and Surfacing system when the potential for a non-analytic intersection occurs.

The Kernel ist designed for extensibility. For thus purposes, it is important that the internals of the Kernel be open to the outside. An object-oriented language is an excellent vehicle for such capabilities. If the Kernel is written in an object-oriented programming language, all topology, geometry, attributes, and much more can be represented as objects. Data and methods for objects can be made accessible to an application so that application-specific data can be added, as can new or alternative methods for manipulating the data.

Application Procedural Interface (API): The Application Procedural Interface encapsulates Kernel functions and is a procedural interface from the application to the Kernel or to the Parametric Curve and Surface system. It presents interfaces for curve, surface and solids construction, sweeping, extruding, Boolean operations, intersections and so on.

These main aspects are fulfilled by the ACIS Geometric Modeller /52/, which will be applied more and more in user-friendly CAD/CAM systems.

7.2.2.2. Requirements related to specific SFM techniques

It is helpful if, during or immediately after the geometric modelling, the user decides the best possible position of the object for the building procedure. Then the application engineer for the selected SFM equipment has only to realize horizontal or vertical translations at the platform.

For this complex problem, some factors which have to be considered are, for example:

- the part structure (surface foundation)
- the main measurements (the maximum height determines the building time)

- the accuracy requirements (essential for shrinkage factors in relation to main measurements)
- the surface quality (layering by the slice procedures; arrangement of support structures)

For the exact definition of selected requirements, a special object description sheet is necessary.

For a systematic evaluation of these different requirements and for mastering them, application of a suitable classification system is necessary (see Section 7.3).

Related to such a classification system, the development of a specific data bank and knowledge base would be useful.

7.2.3. Data handling and interfacing aspects

It has to be recognized, that 3-D geometrical objects have a large data volume. This data volume will be further increased, if the data has to be transferred into a specific data format, as required by the interface.

For most SFM procedures, the STL Interface is used. Here, STL stands for Stereolithography, and means that the first interface developed by 3-D Systems is effectively an industry standard.

Some other partly accepted interfaces, like IGES or CAD system-specific solutions (e.g. CATIA) have to be translated by computer internally into the same STL data format. The STL interface is related to the FEM methodology. The single object surfaces will be subdivided into a large number of triangle surfaces.

Subsequently the neighbour surfaces of the triangle patches will be fitted together in an interactive approach.

Some problems arise in this context, because the STL format consists of an unordered list of triangular facets without any topological information other than the orientation of each facet /53/.

Consequently, it is not always obvious how to mate neighbouring facets, nor is it a trivial matter to determine which facets are neighbours.

The quality of the approximation can be influenced by selecting a tolerance restriction. For a high surface-quality of the object to be generated, the operation with the smallest possible tolerances is requested. This request also needs to be considered during the 3-D geometrical modelling. The above-mentioned specification of the STL format and the requirements leads to some problems /53/.

Typical geometrical and topological problems that arise when processing facet models include: (1) gaps between facets, such that along a given edge of this gap, there is only a single adjacent facet: (2) more than two facets adjacent to an edge i.e., a non-manifold edge; (3) intersections of facets due to self-inter-secting solids, or intersections of multiple solids. (If the intersection occurs along an edge, then it becomes a non-manifold edge.); (4) orientation of the facets consistent with either the vertex orientation (right-hand rule), the explicit facet normal, or with neighbouring facets.

Most of the problems encountered with the STL file format are due to a lack of topological information.

To overcome such problems a new file format, called the RPI format, has been developed (see /54/ and /55/).

In general, a large amount of data has to be handled.

For this purpose, two standard data formats for the physical transmission are possible:

(a) ASCII File (American Standard Code for Information Interchange)

(b) Binary File.

The data format will be determined by the user of the CAD system.

The important advantage of the ASCII format is the readability by text editors or word processors. But it suffers from three times higher data volume, in comparison with the binary format.

The data transfer from the CAD workstation to the SFM equipment can be realized by:

- diskettes (for relatively simple parts, and a small amount of data)
- streamer (nearly unlimited amount of data)
- optical disks (unlimited amount of data)
- networks (unlimited amount of data)

7.3 Process Planning and Operational Planning

Like other CAP processes, for SFM procedures also Process and Operational Planning is the pivot in the entire process chain.

Because of the sophisticated process and many influences, the human factors are of high importance.

7.3.1 General Requirements Concerning Systematic Operational Planning

The general task of operational planning consists of transforming geometrical information into specific process information for building real objects in a reliable way.

The general SFM process sequence comprises

- transferring input data
- checking these data
- making evaluations concerning producibility
- preparing building parameters for positioning and stepwise manufacturing in the light of available possibilities (equipments) and any specific restrictions required by accompanying documentation.

Most customers deliver special recommendations for specific building parameters, shrinkage factors, etc.

To make the entire process chain transparent and reproducible, some additonal information is necessary.

In general, the following aspects have to be considered:

- in agreement with European and International (ISO) Quality-standards, every element and all requirements - especially for process planning - have to be documented in a systematic and orderly manner in the form of written policies and procedures (ISO 9000)
- the application engineer, who is responsible for the further input-data preparation (starting from 3-D data), has to oversee the realization of the entire process. Because of the many influencing factors, he needs some systematic support

- the CAD designer who intends to use SFM techniques also needs at least some basic principles, which have to be considered in the design phase. This aspect is related to the requirements of Concurrent or Simultaneous Engineering, in connection with necessary information feedback about producibility

- every interested customer has to know what kind of processes are suitable for specific part requirements. It would therefore be useful if suitable information is available for specific benchmark tests for evaluation of different SFM systems, depending on the part or factory-specific requirements

For all the requirements mentioned, process documentation and suitable part classification are necessary.

7.3.2 Process documentation for SFM procedures

7.3.2.1 Necessities and general tasks

The efficient utilization of SFM equipment to produce good results requires ordinary process documentation. In most companies a process logbook is used, but the information contained is currently sparse and insufficiently complete for an exact evaluation of process results.

On the other hand it has to be emphasized that, especially for large, thin-walled parts with high accuracy requirements, more than one attempt is normally needed for correct part building (sometimes four or five!). For a quality, cost and time evaluation, the entire process chain has to be considered:

- identification of input data, media, responsible persons

- determination of the geometrical objects, the building possibilities, spatial assortments, arrangements with other parts, necessary support structures, slice parameters

- determination of technological parameters (cure depth, laser power and other physical parameters)

- software-handling, user-support, dialogue mode, graphical aids

7.3.2.2 Proposal for unified process documentation

An operational planning system which can be used for all SFM techniques is needed. This need can be met by development of a general operational register document, providing specific recommendations appropriate to the requirements of the different procedures.

This register document can be subdivided into four main parts

- a form for determination of the object information
- a form for recording operations to be performed
- a form for compiling detailed process parameters
- a summary of time and cost sheet

The third form (process parameters) has to be specific to the system and material. Good examples of such forms are given in DuPont's user manual (Reference /55/). The crux of the proposed general operational sheet I is its relation to the various aspects of FMEA (failure mode and effect analysis). This means that in addition to the necessary object-specific facts which are recorded in the document, it is very important that all essential comments and an evaluation of the results are also recorded /57,58/.

This is demonstrated by an example of a specific challenge test part (compare Chapter 11).

7.3.2.3 Utilization of the information content of the general operational document

There are a variety of different possibilities for utilization of the information content:

- Basis for exact time and cost calculation for every object manufactured by the particular SFM equipment.
- Basis for time and cost estimation for carefully selected parts (based on the proposed part classification and part analysis at the particular company)
- Using of these time and cost elements as a realistic basis for qualified benchmark tests
- Basis for knowledge and rule acquisition in the frame related to the given part class, including the opportunity for continuous qualification of the rule system ("knowledge source").

7.3.4 Postprocessing

The postprocessing tasks are the last in the entire process chain, but are as important as the others.

Because at this stage some manual operations are necessary, the danger of damaging a part is particularly high. Therefore the operator for this last process step has a high responsibility for the successful process realization.

He therefore needs clear instructions, especially concerning the specific part geometry.

These general requirements have to be considered for nearly all types of SFM procedures (except the FDM process).

The necessary postprocessing tasks for some major SFM procedures are shown in Fig. 7.4.

	STL (Stereolithography)	SLS (Selective Laser Sintering)	FDM (Fused Deposit Modeling)	LOM (Laminated Object Manufacturing)
Cleaning	X	X	-	X
Postcuring	X	-	-	-
Finishing	X	X	X	X

Fig. 7.4. Essential postprocessing tasks for different SFM processes

As demonstrated in this figure, the Stereolithography procedures require the highest expenditure for postprocessing tasks.

The requirements are explained in the chapters that deal with the process sequences.

Most important is the fact that, for safety reasons, specific recommendations for postprocessing tasks have to be prepared, especially for cleaning of Stereolithography parts.

8. Application Areas

8.1 General aspects

The first developments of SFM technologies were mainly oriented toward so-called "touch-and-feel" models. This first application range remains are of interest.

But the dynamic developments in recent years have led to many different application possibilities.

A general overview is given in Fig. 8.1.

DESIGN-MODELS	FUNCTION-MODELS	MANUFACTURING-MODELS	FINISHING PROCESS POSSIBILITIES
• CAD-model verification (Design-specification) • Visualizing objects • Proof of Concept • Marketing objects ("show and tell")	• Form- and Fit-models • Flow analysis • Analysis of stress distribution • Mock-up parts (especially in airplane and automotive industries) • Pre-series parts	• Plastic mold parts - vacuum casting - metal spraying • Casting - sand-casting - investment-casting - die-casting • EDM-electrodes	• Cutting - milling - boring - grinding • Sand blasting • Polishing • Coating • Painting (varnishing)

Fig. 8.1. Essential application areas of SFM-objects

All of these different application areas contribute to drastic reductions of the "Time-To-Market" (TTM) for new products./28/

In most cases the objects generated can be used for different application areas. So it can be emphasized that results SFM procedures really are tangible aids for the requirements of Concurrent Engineering.

All these possibilities contribute to another important goal - to improve the quality in an entire company.

The purchase of two 3-D SLA 250 systems at Chrysler was funded by the "Accelerated Quality Improvement Program (AQIP) in 1989. The proposal was directed to improving vehicle quality and reducing cycle time in connection with "Process Driven Design".

Important aspects from this point of view are /29/:

- Allowing quick production of prototype parts and thence a finished part, bypassing the traditional methods where some points, errors, interpretations, machining, an inadequately cut sections delay the construction of a correctly toleranced prototype part or model
- Allowing the discovery of mistakes faster so that the program, pilot, and durability vehicles will be more representative of the first production intent.
- Allowing evaluation of assembly techniques with "Process Driven Design" representative parts relating to fit/finish issues, in less time than current practice
- Allowing the manufacture of patterns and die set combinations for sand casting and moulding operations, thus improving prototype delivery time (i.e., engine mounts, intake and exhaust manifolds, etc.)
- Allowing the building of accurate plastic prototype parts as required
- Allowing the manufacture of complex prototype rubber parts (i.e., shift boots) by generating the tooling for a rubber mould

All these important aspects normally also lead to drastic cost savings, in comparison with traditional procedures (see Chapter 11).

8.2 Design support

As shown in Fig. 8.1, some different aspects in this working area have to be considered:

- CAD model verification
- Visualizing objects
- Proof of concepts
- Marketing objects ("show and tell"-objects)

8.2.1 CAD model verification

In theory, that every designer needs to generate physical objects for confirmation of his own results. This is especially necessary for aesthetically designed parts. Another important aspect of CAD model verification in connection with the product development is behaviour analysis. The CAD model can be used, to generate a finite- element mesh or for stress analysis, to generate tool design, and other purposes.

8.2.2 Visual objects

The necessity of visualizing objects is characterized by T. Mueller as follows. "The simple fact is that many people cannot visualize a part by looking at a print. Even engineers and toolmakers who deal with prints every day require several minutes or even hours of studying a print to understand a complex part. Unfortunately many of the people who approve a design, a group which typically includes senior management, marketing analysts, and customers, have much less ability than engineers and toolmakers to understand a design by looking at a drawing. Prints simply cannot communicate a design well to all people. Consequently, there is always the risk that design decisions will be made on the basis of an inaccurate understanding of the design. Such decisions are often very costly down the road" /28/

The "proof-of-concepts" aspects relate especially to adaptation of specific details to an entire object environment.

As an example, the injection moulded shielding of a car telephone in its storage box can be mentioned. Apart from the "form and fit" functions, which are mainly connected with the sizes and measurements, the "proof of concept" is related with aesthetic aspects in the entire interior environment of the specific car.

The aspects of "marketing objects" can be understood in the sense of visualizing. Everyone can assess a project better if an object model is available in comparison with the usual sketch, scheme or drawing. In other cases, a physical object is useful for qualified decision-making.

The possibility for qualified customer discussions based on visualized objects can be classified in the same application range.

8.3 Function models

Based on the improved performance of materials now available for SFM technologies, some possibilities for functional models are given below.

8.3.1 Packaging models

The SFM techniques allow eas scaling down (or up) of the size of a model, or choice of measurements.

In case of limited space, for example in the engine compartment of an automobile, a specific object can be adapted. An example could be the water tank in a car. Varying the sizes by way of 5, 10 or 15 % reductions, or if possible 5, 10 or 15 % extensions, the best suited part can be determined. All CAD data can be changed accordingly and used for all other design and process preparation tasks.

8.3.2. Form and Fit models

Typical application areas for form-and-fit models are related to uses in

- the automotive industries
- aereplanes
- consumer products

As well as the main packaging aspects connected with sizes or volumes, the forms also have to be considered from the aesthetic and functional aspects. For example the mirror housing of a new car design has to be evaluated and adapted from the form to the general outside design. This includes in every case the fitting to the door. Other examples for aesthetically designed forms and fitting to the general environment are

- storage boxes or,
- the car telephone-holder

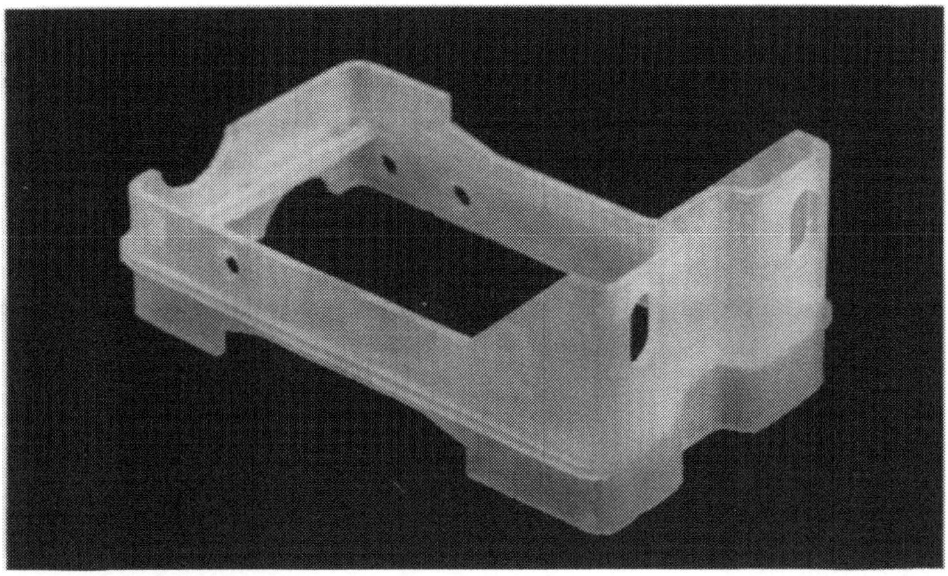

Fig. 8.2. Car telephone-holder (BMW)

Further examples are the housings for consumer goods like

- cameras
- videorecorders
- kitchen equipment

In all cases, advanced outside shape designs, and the exact fitting of different housing elements are requested.

8.3.3 Flow analysis

Flow analysis is necessary for specific objects in the

- aerospace / aeroplane
- automotive, or
- shipbuilding industries, and others

Traditional manufactured objects for air flow, water or other fluid tests cannot be changed and adapted by computer support. But if the 3-D original data are stored in a computer model, every changing of object data based on specific tests can be realized with computer support. A changed model can be generated again by SFM techniques much faster than in traditional ways. Flow analyses are also useful for inner parts like inlet manifolds, exhaust pipes or similar. An example is shown in fig. 8.3.

Fig. 8.3. Air blowing wheel (BMW)

8.3.4 Analysis of stress distribution

For the stress analysis using photo-optical methods, physical objects are necessary. If the material features of the SFM objects are nearly identical to those of the real functional parts, the can be used for determination of the stress distribution.

8.3.5 Mock-up parts

The term "mock-up", first introduced in the aircraft industry, is used for final testing of parts from different and all necessary points of view. Normally, mock-up parts are assembled in the complete product and functionally tested for a pre-determined time.

8.3.6 Pre-production-parts

In cases where mass-production will be introduced, pre-series production of 10,20 or more objects is usual. The pre-series parts are especially used for further process preparation, which means that with the pre-series-parts all necessary fixtures, chucks, special tools and measurement devices are developed, or provided, and checked.

8.4 Manufacturing models

The manufacturing models can be characterized as the bridge between production preparation and realization.

The manufacturing models, generated by SFM-technologies can be used for transforming the geometrical pattern into other materials.

Most important application ranges are

- plastic mould parts
- casting
- EDM electrods

8.4.1 Plastic mould parts

The main idea for this application is the possibility to produce multiple copies of a prototype, with greater durability.

8.4.1.1 Vacuum casting with Silicon Moulding

Silicon moulding compounds, and their use as a tooling aid, have been available to the prototype plastic and plaster mouldmaker for many years /30/.

The following advantages can be emphasized:

- an extremely high resolution of master model detail can be copied to the silicone cavity mould

- reduction of backdraft problems (i.e. die lock, or the inability to release the part from the mould cavity because some of the geometry is not within the same draw direction as for the rest of the part)

These problems are commonly encountered using hard moulds, making it necessary to have loose pieces, which are virtually eliminated when using the silicone moulding process. Silicone moulding compounds can offer superior master model detail complexity when compared to traditional cold synthetic rubber moulding material. Also, there is an advantage over cold moulding rubber material with regard to the number of copies that can be made out of a single mould (of the order of 15 percent more copies may be accurately produced without degradation of the part tolerances).

Fig. 8.4. Example for vacuum casting

When creating moulds using silicone moulding material there is no need to use parting agents or materials like Polyvinyl Acetate (PVA), because the silicone material itself has superior release characteristics.

Because of this advantage, it is not necessary to spray PVA onto the master model and wait for it to dry between applications (usually two are required). A ten percent time advantage is realized when compared to traditional hard moulding techniques.

8.4.1.2 Metal spraying

Using metal spraying, it is possible to create very quickly an injection mould that can be used to mould a limited number of prototype parts.

The metal spraying process is operated manually, with a hand -held gun. An electric arc is introduced between two wires, which melts the wires into tiny droplets. Compressed air blows out the droplets in small layers of approximately

0.5 mm of metal.

The droplets cool rapidly and are nearly solidified by the time they reach the work surface, so that they do not immediately burn into the work surfaces. An example for making a mould with the MCP/TAFA metal arc spray system is shown in Fig. 8.6.

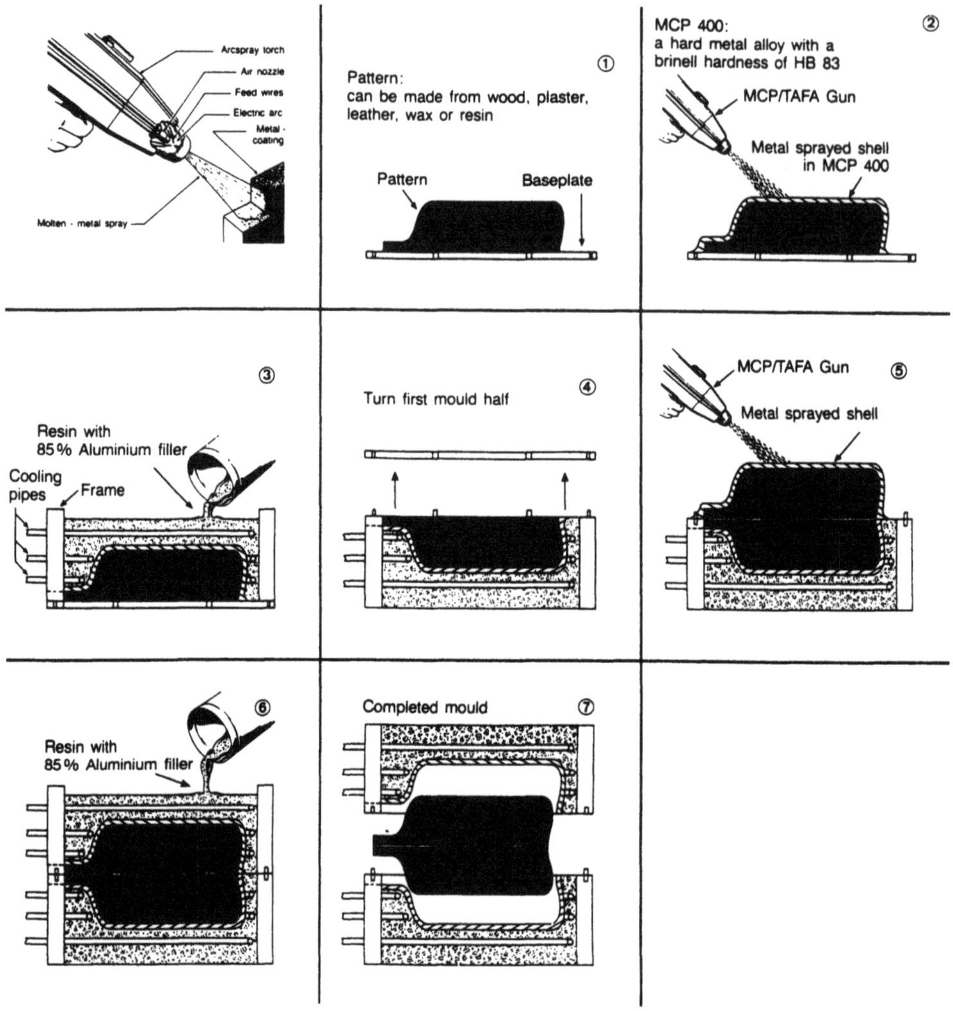

Fig. 8.5. MCP/TAFA Metal arc spray system

The technology for metal spray toolmaking is not particularly new, but its use in the investment casting industry is becoming more practical, thanks to the increased use of SFM techniques.

The significant advantages can be characterized as follows:

- this tool/mould making technique is a process for reproducing shapes very accurately (reproduction is so accurate that the metal shell is comparable to an electroform in its precision - but much quicker!)

- The process is versatile and not limited for size

- the process is simple and straightforward
- it reduces cost by as much as 90 percent
- it provides closer control of the tooling process
- it requires only a modest equipment investment
- it makes possible the fabrication of a part for final design and approval in the minimum elapsed time

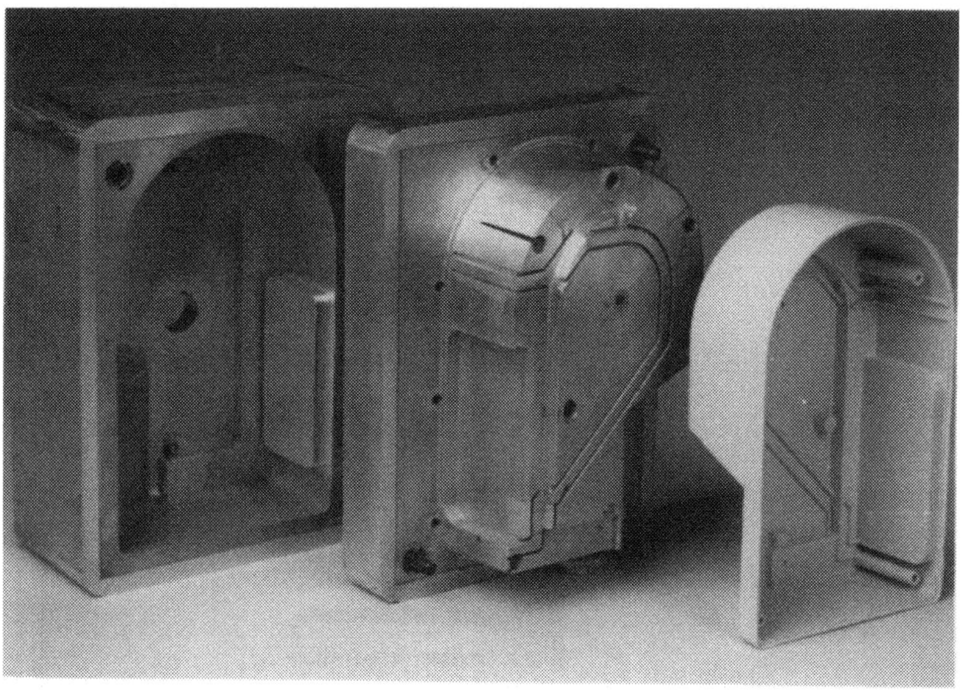

Fig. 8.6. Example for metal spray

For limited production needs the moulds can be used for 1000 - 2000 injections, which is in many cases enough for production of hardware. So the metal spray toolmaking can be used in connection with SFM techniques as a complete process chain for small and medium sized batch production.

8.4.2 Casting /30/

Metal casting plays a major role in the entire manufacturing field. In the metal casting process a metal, or more commonly an alloy, is heated until it is molten, whereupon it is poured into a mould or die that contains a cavity which represents the shape of the component or casting to be produced.

Moulds are invariably expendable, being destroyed when the casting is removed.

Dies are permanent, and may be used to produce a succession of castings.

Patterns may also be permanent or expendable. A permanent pattern is re-usable, whereas an expendable pattern is consumed during the process of mould or casting production.

An explanation of the principle of moulds and patterns is given in Fig. 8.7.

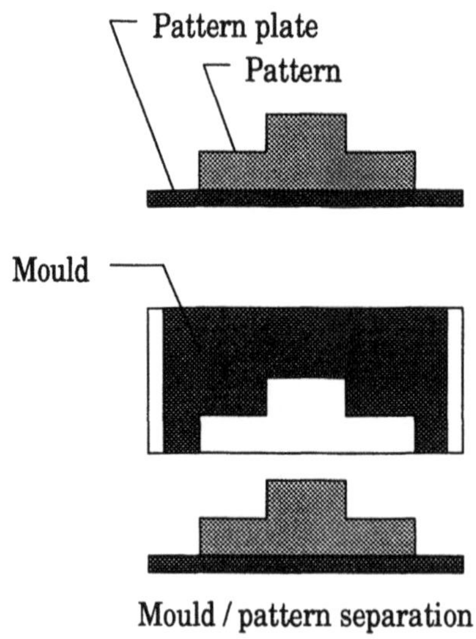

Fig. 8.7. Mould and pattern

It should be pointed out that the objects generated by SFM processes are suitable for different casting processes (see Table 8.1).

Permanent Pattern/Expendable Mould Processes

- Sand Casting Processes
  - Green sand moulding process
  - Dry sand moulding process
  - CO_2/Silicate process
  - Shell moulding process
  - Cold set processes
  - Vacuum-sealed moulding process
- Refractory Aggregate Processes
  - Ceramic mould processes
  - Plaster mould process

Expendable Pattern/Expendable Mould Processes

- Sand Casting Process
  - Evaporative pattern casting process
- Refractory Aggregate Process
  - Investment casting process

Permanent Mould Processes

- Die Casting Processes
  - Gravity die casting
  - Pressure die casting
  - Low-pressure die casting
- Squeeze Casting
- Counter Gravity Casting
- Centrifugal Casting
- Continous Casting

Table 8.1: Classification of Casting Processes

Because of the broad application of sand-casting for this process, most of the applications of SFM objects are realized. The general range of casting alloys and principal casting processes are shown in Table 8.2 /30/.

Alloy Group	Specific Alloy	Principal Casting Processes
CAST IRONS	Grey cast irons (flake graphite) Nodular irons (spheroidal graphite) Malleable cast irons Heat resistant, chemical resistant and wear resistant alloy irons	sand casting
STEELS	Plain carbon steels Low alloy steels High alloy steels Heat resistant, chemical resistant and wear resistant alloy steels	sand casting ceramic mould process investment casting
ALUMINIUM	-	sand casting die casting investment casting
COPPER	Brass Bronze Gunmetals Aluminium Bronze Silicon Bronze	sand casting die casting
MAGNESIUM	-	sand casting die casting
ZINC	-	pressure die casting
SUPERALLOYS	Nickel-base Cobalt-base	investment casting
MISCELLANEOUS	Titanium alloys Bearing metal alloys Magnet alloys Refractory alloys	-

Table 8.2: The Range of Casting Alloys

The Investment Casting Process, also known as the lost-wax process, is probably the process most commonly associated with the term precision casting.

The Die Casting Processes employ a permanent, re-usable mould or die, generally made of metal. This feature distinguishes them from the refractory aggregate moulding processes considered so far. The die casting processes are normally precision casting processes, of four kinds:

- Gravity Die Casting (permanent moulding)
- Pressure Die Casting (die casting)
- Low Pressure Die Casting
- Squeeze Casting

For further explanations see /30/

A most important point for consideration is the magnitude o the solidification contraction, in comparison with the shrinkage factors at SFM processes. The solidification shrinkage factors for selected metals are shown in Table 8.3

Metal %	Volume Contraction
Iron	3.5
Nickel	4.5
Copper	4.2
Aluminium	6.5
Magnesium	4.1
Zinc	4.7
Lead	3.5
Tin	2.3

Table 8.3 Solidification shrinkage

8.4.3 EDM electrodes

A method successfully tested in research laboratories but so far not widely applied in industry is the possible manufacturing of copper electrodes for EDM processes (EDM - Electro-Discharge Machining). To create the electrode, the SFM objects are used to create a master for the electrode. An abrading die is created from the master by making a cast using an epoxy resin with an abrasive component. The resulting die is then used to abrade the electrode. A specific advantage of the SLS procedure (see Section 6.1) is the possible usage of other materials. Using copper in the SLS Process, it is possible quickly and affordably to generate the electrodes used in electrode EDM.

8.4.4 Master models

Not so far broadly applied, but possible in principle, is the application of so-called master models. Such models can be used, especially for copy milling, if the subcontractor for the required parts is a small company without CNC machines. On the other hand, the principle of generating master models only when necessary, allows some storage space to be saved.

Normally, in traditional manufacturing processes, the model- maker or pattern-maker needs a lot of storage space for the normally used wooden dies and patterns.

8.5 Finishing process possibilities

Because of such various influencing factors as shrinkage, distortion, curling and accessible surface smoothness, it is necessary and useful to apply some additional procedures. Furthermore, additional procedures are necessary in specific cases, for example when manufacturing threads, which is normally not possible with SFM procedures.

8.5.1 Cutting processes

The most frequently used resins and other materials allow the additional application of cutting processes.

Such typical cutting processes as

- milling
- boring
- turning
- grinding

are especially necessary in the following cases:

- deviations in distance measurements or tolerances due to difficult shrinkage calculations (depending on the materials and part structure)
- incomplete generation of selected form features caused by process preparation which is difficult to oversee
- in the case of necessary supports or other remainders of materials, some inaccuracies arise. It is difficult to correct this manually in an economic and quality-justified manner.

In all these typical cases it is possible to achieve economic finish machining of the objects generated with a combination of NC machining and computer-aided NC programming.

8.5.2 Sand-blasting and polishing

Sand-blasting can be used more or less as an additional cleaning operation. So it can achieve a better surface quality, but not an increased accuracy.

If a smooth surface is requested, it can be realized by polishing after sand-blasting.

8.5.3 Coating

The coating of plastic geometric objects can be used for further improvements of the physical properties of the surface

One example could be galvano-coating, which provides very thin metallic layers.

8.5.4 Painting

Painting is mainly used for improved appearance, for example for marketing objects.

8.6 Comparision with other manufacturing processes

With currently available machine tools, machining requirements of given workpiece assortments can be fulfilled mainly by milling; if there are holes included, also by drilling.

Essential criteria for comparision of SFM technologies with traditional manufacturing processes are given as follows:

- materials (as the current main difference)
- overall measurements (complexity)
- part structures
- accuracy

A rough characterization of typical and respectively possible application ranges are shown in Fig. 8.8.

Material	Size	Part Structure	Accuracy (Tolerance)	
- Plastic - Nylon - Vax	small	Complex Prismatic (Sculptor Surface)	low < 1 mm	typical
- Copper Aloys - Ceramics - Aluminium - Iron (Indirect)	medium < 1 m^3	Simple Prismatic	average ~ $\frac{1}{10}$ mm	possible
- Steel	large > 1 m^3	Simple (Sheet metal, rotational parts, etc.)	high $\frac{1}{1000}$ mm	impossible

Fig. 8.8. Application ranges and currently given restrictions for SFM-technologies

Relative to part structures, some typical features for advanced direct model generation by SFM technologies in comparision with NC-milling are as follows:

- small and thin (unstable) parts
- outside and inside shapes and shape elements
- hollow parts, tubes
- parts which can be difficult to clamp

In general it can be emphasized that for all types of manufacture of freeform parts, lead times obtained by application of SFM procedures are drastically reduced in comparision with NC-milling.

An exact comparision with other manufacturing processes for detailed evaluation requires the inclusion of all essential cost and time factors by consideration of the entire process chain.

9. Manufacturing Materials

The different application areas described are closely related to the properties of the different usable materials. The materials currently used show marked differences from traditional materials in manufacturing industry. As in Chapter 3, a subdivision can be made into three areas:

- fluids
- powder for solidification
- wire and laminates

For all these different materials, varios factors and relationships have to be considered, as shown in Fig. 9.1 (in agreement with /33/). All five of these factors are interrelated and changes in one will, in general, influence the others. One of the elements, however, has a strong interaction with all of the other four variables. This dominant system element arieses from the characteristics of the photosensitive materials used in the Stereolithography process /33/.

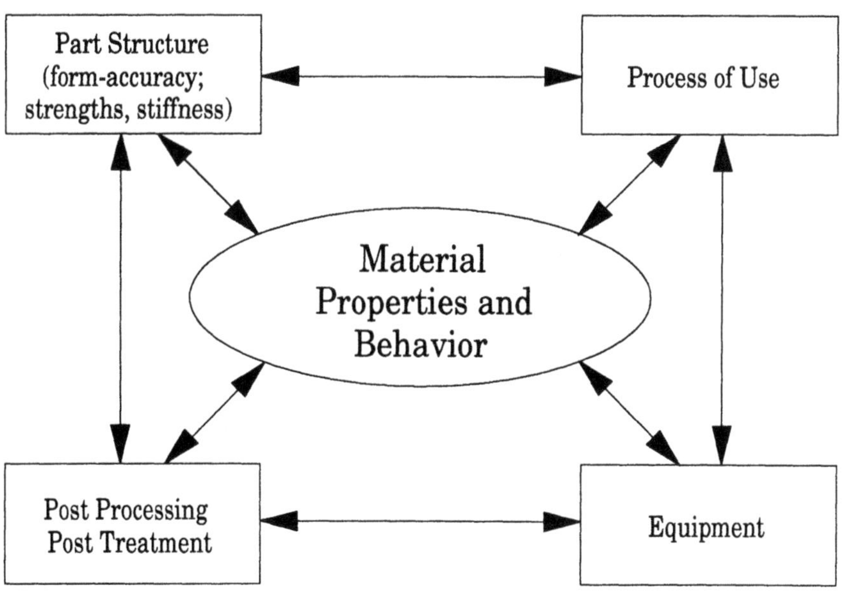

Fig. 9.1. Influence factor and relationships concerning material properties and behavior

9.1 Fluids

It has to be emphasized that photopolymer materials, especially, are highly sensitive and influenced by several factors. For a detailed explanation of the actinic processes and their theoretical foundation, specific references are available /13, 33/.

Therefore this book will stress the application aspects, mainly related to Stereolithography.

The following characteristics are important /33/.

- Physical properties
- Basic material characteristics
- Material handling procedures
- Postprocessing procedures
- Health and environmental aspects
- Imaging properties

9.1.1 Physical Properties

From the application point of view, the final physics properties are the essential factor.

The solidified phase can be characterized by different features, as shown in Table 9.1.

These features are typical of the ranges of other traditional thermoplastic materials, and logically very far from typical metallic materials. This is currently the main restriction limiting broad universal use in the fields of mechanical engineering. On the other hand, the main features allow a relatively broad usage in all application ranges characterized in Fig. 8.1. All the significant material properties are under continuous development. Furthermore, it appears that, leading companies in this field are able to adapt the material properties to customer requirements. In Table 9.1, the material properties of some resins for Stereolithography are included.

UV-Laser Cured "Green" Part (Typical)

Tensile Strength	16 MPa	170 Kgf/cm²	2400 lb/in²
Elongation at break	15 %	15 %	15 %
0,2 % Offset yield	6.9 MPa	71 Kgf/cm²	1000 lb/in²
Secant modulus at 1 % strain	460 Mpa	4700 Kgf/cm²	65000 lb/in²
Flexural Strength		[material did not fail]	
Yield Strength at 5 % strain	10 MPa	100 Kgf/cm²	1500 lb/in²
Secant modulus at 1 % strain	310 MPa	3200 Kgf/cm²	45000 lb/in²
Secant modulus at 5 % strain	200 MPa	2100 Kgf/cm²	30000 lb/in²
Hardness (Shore "D" durometer)	77		
Izod Impact (notched)	14 KJ/m²	14 J/m	0.26 ft-lb/in
Density	1.205 g/cm³		

Post-Cured by fluoreszent or UV light (Typical)

Tensile Strength	21 MPa	220 Kgf/cm²	3100 lb/in²
Elongation at break	9.2 %	9.2 %	9.2 %
0,2 % Offset yield	13 MPa	130 Kgf/cm²	1800 lb/in²
Secant modulus at 1 % strain	810 Mpa	8200 Kgf/cm²	120000 lb/in²
Flexural Strength		[material did not fail]	
Yield Strength at 5 % strain	20 MPa	210 Kgf/cm²	3000 lb/in²
Secant modulus at 1 % strain	650 MPa	6600 Kgf/cm²	94000 lb/in²
Secant modulus at 5 % strain	420 MPa	4300 Kgf/cm²	61000 lb/in²
Hardness (Shore "D" durometer)	80		
Izod Impact (notched)	15 KJ/m²	15 J/m	0.28 ft-lb/in
Density	1.205 g/cm³		

Table 9.1: Example of material properties of the SOMOS-resin 3100 /33/

9.1.2 Basic material characteristics

These features are concerned with the fluid material stage

9.1.2.1 Viscosity /33/

The viscosity, (the ability to flow) of Stereolithographic resin mixtures influences mainly the uniformity, and thus the accuracy, of the "recoating process". To a lesser extent, it also influences, the ultimate shrinkage in the part. In general, these two factors are opposite in their impact on the imaged model. With some machine designs, the most accurate application of photopolymer layers is obtained at lower viscosities (viscosity < 750 cP).

To achieve these low viscosities, the resin vendors have incorporated into the compositions a large amount of low molecular weight monomers. Since the amount of shrinkage in a photopolymer is directly related to the density of potentially "reactive sites" in the formulation (more are present in a formulation with a higher concentration of low molecular weight monomers), more shrinkage, and its accompanying distortion, are usually present in lower viscosity resin mixtures. In contrast, at solution viscosities > 8,000 cP the length of time required to achieve uniform layer thickness can be unacceptably long. Photosensitive materials at these higher viscosities, however, usually exhibit fewer accuracy errors induced by shrinkage. New "recoater" designs have been proposed in patents and patent applications that offer the ability to obtain improved layer accuracy and that advantage of the reduced shrinkage usually associated with high viscosity resin mixtures.

9.1.2.2 Surface Tension

The surface tension, affecting the ability of the liquid resin mixture to recoat an imaged layer uniformly, can have an influence on the reproducibility of the Stereolithography process. If the resin mixture does not sufficiently "wet" the previously imaged layer , then an irregular thickness will result from the recoating process. Moreover if the surface tension of the liquid resin is too high, it is possible that when recoated, the resin mixture will tend to exhibit increasing thickness from the edge to the interior of the previously imaged layer (i.e., a meniscus effect). Most photosensitive compositions used in Stereolithography do not show these potential defects.

9.1.2.3. Volatility

The volatility of components in the resin mixture can have a detrimental impact on system accuracy and operator safety.

It has been reported that volatile reactive ingredients can vaporize and condense on some of the interior components of the Stereolithography machine. This is especially critical if these reactive materials coat the optical system and disrupt the imaging process. High volatility of ingredients can also lead to increased odour and greater potential for operator contact.

9.1.2.4 Material Uniformity and Stability

The uniformity and stability of the photopolymer mixture in the vat can influence the accuracy and the physical properties of the imaged part. If separation of the mixture occurs during the process of making a part, it is possible that the photosensitive material can show different light sensitivity, and hence different accuracy, during the process. Most commercial resins are stable and do not exhibit separation.

9.1.3 Material Handling Procedures /33/

9.1.3.1 Vat Material Temperature

Control of the photopolymer temperature during the Stereolithography process is important for the production of accurate models. As the temperature is increased the solution viscosity usually decreases. If the temperature of the resin in the vat varies during the part manufacturing process, it is possible to introduce some inaccuracy in the individual layer thickness. Most Stereolithography systems control the temperature of the photosensitive resin in the imaging vat.

9.1.3.2 Bubble Formation/Release

With some part designs, resin mixtures, and process conditions it is possible to generate small bubbles in the vat. If these bubbles are trapped in the resin during the imaging process, non-uniformity and inaccuracy can result. It is also desirable for the surface tension and viscosity of the photosensitive mixture to be low enough to permit bubbles, if formed, to move to the surface and break. The formation of bubbles depends also upon the recoating method.

9.1.3.3 Addition of New Material to the Vat

During the Stereolithography process, fresh material is added to the vat to compensate for that consumed while making the part. It is important for the resin to be a robust formulation so that small changes that might occur with the material in the vat are insignificant relative to the fresh material added. Most commercial materials are sufficiently uniform and robust when used according to the recommended process conditions.

9.1.3.4 Contamination

Contamination of the photosensitive mixture should be avoided, in order to maintain the desired performance. Small amounts of added materials, such as machine oil, dirt, etc., can influence the imaging performance of most resin mixtures. If the characteristic imaging parameters (E_c, D_p) change as a result of contamination, inaccuracy can result due to incorrect exposures being used to image the individual layers. Therefore a clean laboratory atmosphere is required.

9.1.4 Imaging properties

The correct exposure conditions are most important for avoiding errors.

9.1.4.1 Material "Working Curve" /34/

Accurate determination of the variables that define the "working curve" for the resin mixture is probably the most important factor in controlling the accuracy of the Stereolithographic process. In the practical case of Stereolithography, the relationship between the thickness of the exposed layer and the applied energy is called the "working curve".

The equation representing the "working curve" relationship is

$$C_d = D_p * \ln * (\frac{E}{E_c})$$

where C_d is the cure depth or thickness of the solidified layer, D_p is the "penetration depth" or slope [$\Delta C_d / \Delta \ln (E)$] of the working curve, E is the exposure energy used to solidify the resin mixture and E_c is the "critical exposure" at which solidifcation starts to occur.

If the critical parameters (E_C, D_P) that control this curve are not accurately determined and controlled, substantial errors in the cure depth are likely.

In addition to the determination of E_C and D_P, it is important to determine the line width compensation factor accurately for the photosensitive material being used. The line width in the Stereolithography process is, in part, influenced by the amount of light scattering present in the resin mixture. Light scattering is a consequence of certain additives that can be included in the photopolymer to achieve the desired balance of imaging and/or physical properties.

Precise control of the laser beam is therefore very important.

9.1.5 Postprocessing procedures

Essential postprocessing procedures for resins related to the material behaviour are cleaning, post-curing and finishing.

9.1.5.1 Cleaning and swelling

The cleaning procedure requires a good tensile strength for handling the parts (see also Section 4.1.4). The parts are typically cleaned with solvent to remove unreacted photosensitive resin. Depending upon the build principle and the extent of cross-linking in the resin, the imaged part may be distorted during the cleaning process.

9.1.5.2 Post-curing

For some resins, the post-curing procedure can lead to shrinkage and distortion (see Section 4.1.4). Because of such influences, it may if be helpful to realize the post-curing procedure in a vat of water.

9.1.5.3 Finishing

Most available resins allow finishing procedures for smoothing by sand-blasting and surface improvements, and then polishing until a mirror quality is reached.

9.1.6 Health and Environmental aspects

It generally has to be considered that the photopolymers used are toxic materials. Independent of the different properties and the material-dependent recommendations, the following requirements are essential:

- The wearing of plastic gloves, laboratory coats, eye goggles, etc., is necessary for avoiding any skin contacts with resins or "green parts".
- The cleaning equipments (using chemical materials like isopropanol or acetone) and the SFM equipment themselves require exhausters.
- The safe storage of new and old resins has to follow specific recommendations.

Whereas the Stereolithography resins can be used almost completely, the Solider principle (Cubital) does not allow a full utilization of the applied materials. The waste mixture, of resin and wax, cannot be recycled and has to be returned to the vendor.

9.2 Powder, wire and laminates

The materials for all other applied SFM procedures are non-toxic. In particular the FDM procedure (using nylon wires) and the LOM procedure (using Laminates from paper, wood and plastic) do not require any specific expense. Also for the SLS procedure, one advantage of the materials used (Polycarbonate, Nylon, Wax, ABS, SAN) is the non-toxic behaviour.

The essential properties are the following: Flex ModulusFlex strength

- Polycarbonate 120.000 psi 5.000 psi
- Nylon 160.000 psi 8.000 psi

10. Related Manufacturing Process

The major traditional procedures for manufacture of freeform parts are shown in fig. 10.1, which demonstrates that the most important existing procedure for manufacturing of models, patterns, and dies is provided by NC milling. The other related procedures normally employ the widely available milling methods such as

- 3 axis NC milling for typical die-making machine tools
- 3 axis high-speed NC milling
- 5 axis milling

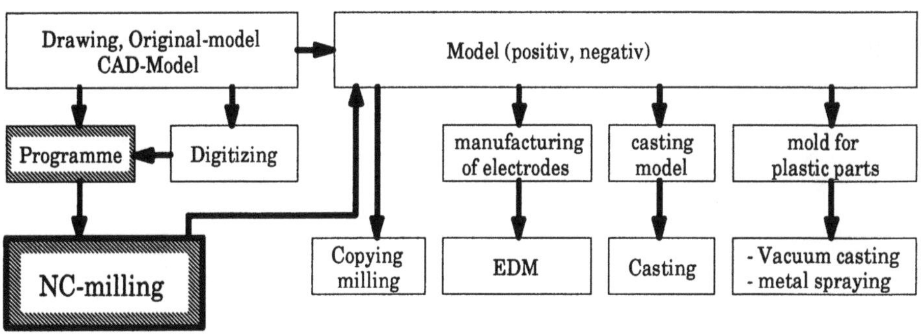

Fig. 10.1. Essential traditional procedures for freeform parts

It is not intended, in this book, to explain the basic principles of NC technology nor the principles of multi-axis milling; both are available in special literature /35,36,37/. In this context, only some specific application aspects related to SFM procedures will be described.

10.1 3-axis NC milling

10.1.1 Application areas

The 3 axis NC milling method is widely available in industry and normally used for

- manufacture of models (wood, plastic, ...)
- manufacture of dies (steel, iron, alloys, ...)

As indicated in Fig. 10.1, the essential process chain can be realized in the following steps:

- Generation of input data
- Manufacturing of an Origin Model
- NC data generation, of different kinds
- Milling of models, patterns or dies

The principal steps are demonstrated in more detail in Fig. 10.2, which shows the high relevance of standardized interfaces in the entire process.

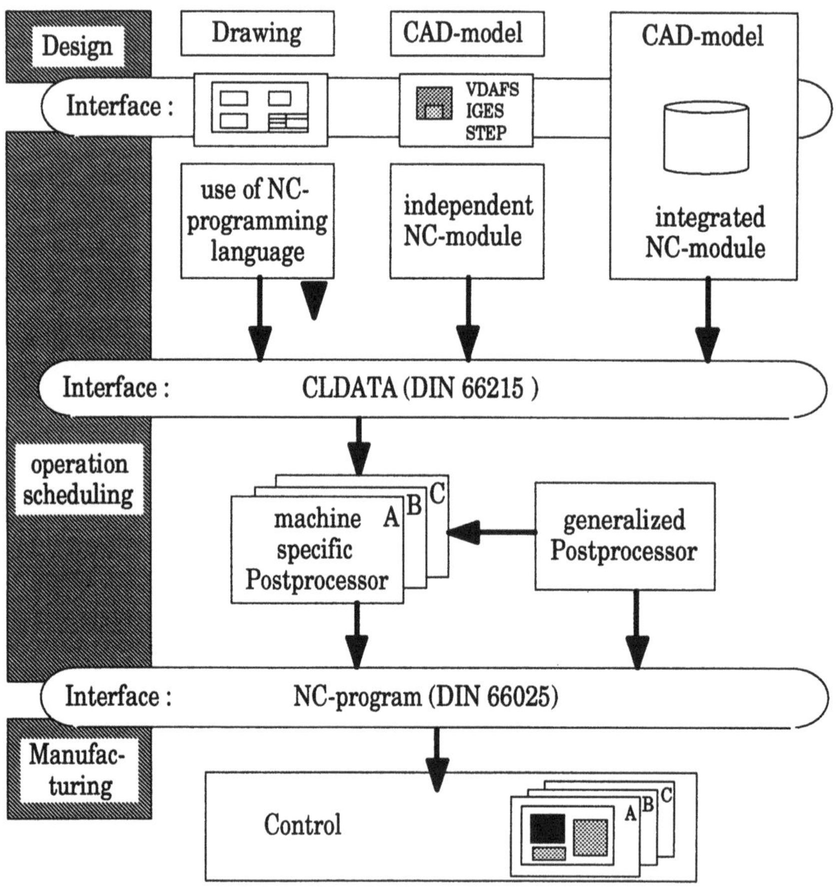

Fig. 10.2. Dataflow from Design to Manufacturing

It shows also the increased usage and application possibilities of digitized data.

The 3-axis milling is closely related to the original and now widely used application of CAD/CAM systems for freeform shape. Relevant well known literature includes /36,38,39/.

10.1.2 NC milling based on digitized data

A relatively user friendly and economical approach to NC milling is provided by the NC data generation based on digitized data.

Depending on the performance of digitizing systems, the digitized data can be used in different ways:

- generation of milling path
- modelling of CAD data
- quality assurance of freeform shapes

The digitizing of freeform shapes, with the aim of generation of milling paths, is one relatively widely applied method.

Typical application ranges are:

- copying milling
- direct milling based on digitized data
- milling based on modified digitized data

For all these possibilities, the following similarities apply:

- the digitized model-points are points of support for the milling path
- the digitizing strategy determines also the milling strategy - the geometrical form of the styling ball must be identical with that of the milling tool
- the diameter of the milling tool can only be smaller or a multiple of the diameter of the styling ball.

Whereas in traditional copying milling, the coupling between the probe procedure and milling is effected by mechanical aids, for both other variants electronic coupling elements are in use. By this means, processing of digitized data is possible without basically changing this data.

Typical modifications of digitized data are

- geometrical data
  - rotation
  - mirroring
  - translation
- conversion of digitized data to the contour surface
- conversion of the contour surface to the centre line path of the milling tool
- magnifying/reducing in size
- positive/negative changings
- compensation of the milling radius

All these calculations are based exclusively on point data. Generation of a surface model which is independent of the digitizing procedure is not provided by any available CAD/CAM system. Such a surface model is necessary for complete decoupling of the model digitizing (model-probe-procedure) and is a prerequisite for universal multi-axis milling, especially five-axis milling. But this is necessary for the most suitable digitizing of any chosen surface and subsequent convenient multi-axis milling operation, with any kind of machine or other tools. With the increased decoupling of digitizing and milling, the possibilities of utilization of digitized data are also increased, but the necessary costs are much higher. This principle is demonstrated in Fig. 10.3.

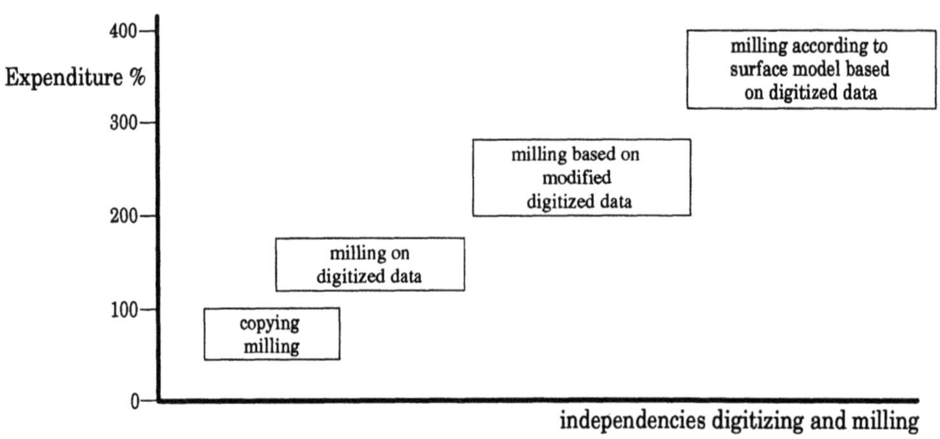

Fig. 10.3. Change of expenditure by decoupling of milling and digitizing-procedure

The utilization of NC milling based on digitized data is an important field of application, but it is more or less restricted to aesthetically designed and relatively simply structured parts, without many different shape elements or form features.

For three-axis milling, the tools used are generally ballend mills. These tools have fulfilled the manufacturing task, not in an ideal kind, because the speed is equal zero in the control position of the milling axe.

In comparison with SFM procedures, the following typical application range can be specified (see Fig. 10.4).

Feature	Stereolithography	3-D milling
Size	< 100 mm^3 (~ 4 $inch^3$)	< 100 mm^3 (~ 4 $inch^3$)
Hole and slot depth / width	< 0.5	> 0.5
Wall thickness / Wall height ratio	< 0.2	> 0.2
Volume object / Volume of bounting box ratio	< 0.5	> 0.5
Number of surfaces in CAD Model	> 100	< 100
Tolerance range	> ± 0.1 mm (~ 0.004 inch)	< ± 0.1 mm (~ 0.004 inch)
Surface finish	R_a > 5 µm (~ 0.2 mil)	R_a > 5 µm (~ 0.2 mil)

Fig. 10.4. Application ranges of NC-milling or stereography for manufacturing of EDM electrodes /33/

10.2 3-axis High-speed cutting

Based on the technical progress of all the essential components

- machine tools
- drives

- tools
- controllers
- spindle bearings etc.

the possibilities for high-speed machining will be increasingly applied. This allows drastically increased speeds.

Cutting speeds up to 3000m/min. and feed rates up to1000mm/min. are possible for high speed-milling. This not only results in a higher rate of stock removal, but also in improved surface finish, less heating of the workpiece, reduced vibration during machining, lower cutting forces and improved tool life.

The HSC machines have been specially designed to provide the higher slide accelerations necessary for precision high- speed milling at feeds up to 10000mm/min. /40/

Graphite spark-erosion electrodes can be produced 10 to 20 times faster by high-speed milling on HSC machines than conventionally produced copper electrodes.

Furthermore, high-speed electro-discharge machining with graphite electrodes halves the time required for spark erosion of moulds and dies.

A typical example is the HSC1200 High-speed machining centre, with tool changing magazine and NC dividing head (see Fig. 10.5).

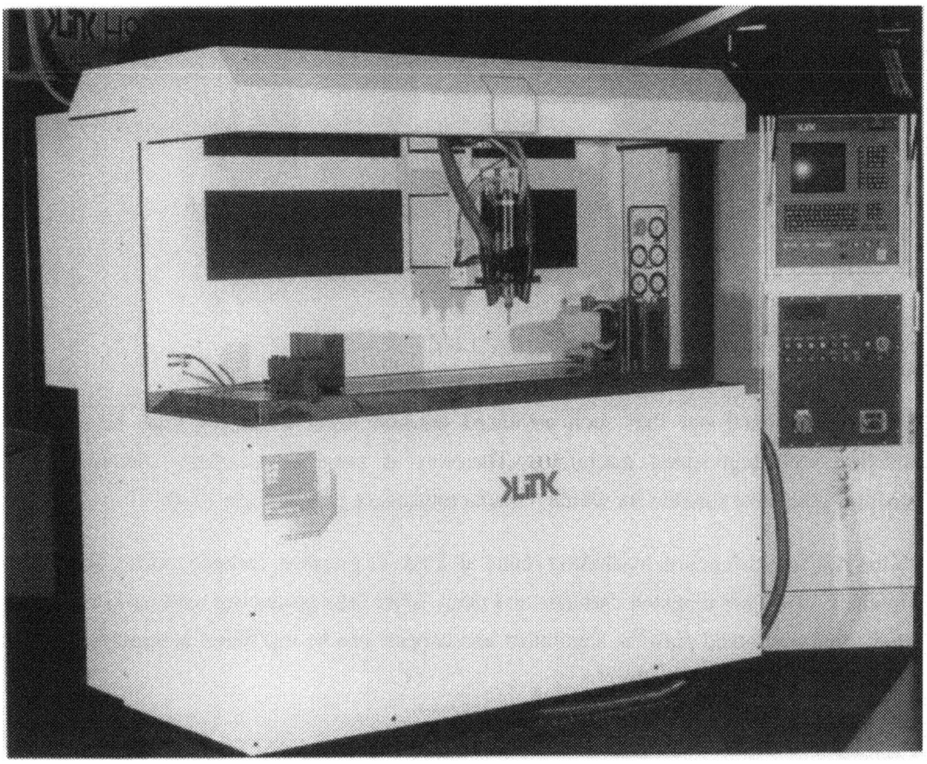

Fig. 10.5 HSC1200

The modular construction enables the HSC high-speed milling machine to be equipped with up to 8 axes. HSC machines can therefore be employed for drilling, 3-D milling, shaping of contours, turning and polishing. An automatic tool changer can be fitted if a wide range of machining operations has to be carried out (see Fig 10.6).

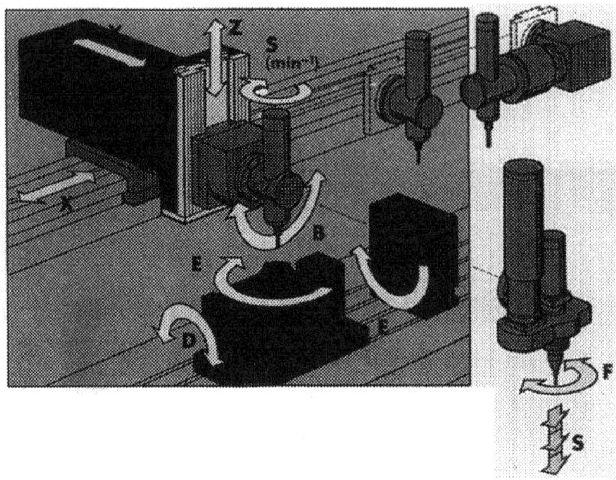

Fig.10.6 Arrangement of CNC axis of HSC 1200

It should be pointed out that such advanced machine tools guarantee high flexibility in connection with high-speed machining. Therefore it must be carefully checked which procedure is the most suitable for which manufacturing task (see Section 10.4).

With the HSC multi-function machining centre all kinds of graphite, ceramic oxides, aluminium compounds, raw state tungsten carbides and many other dust-producing materials, as well as plastics, fibre-reinforced plastics, aluminium and copper, can be machined accurately, without trouble and extremely efficiently.

The machine concept is specially designed for carefree and efficient dust extraction, and features complete protection of slideways, motors, scales and electronics against abrasive and electrically conductive dusts.

10.3 5-axis milling

Five-axis milling is one of the most advanced and highly sophisticated NC Technologies. Therefore NC data preparation for the machining process is one of the most complicated and time-consuming tasks. But in comparison with 3-axis milling, it should be said that manufacturing times can be reduced and surface quality can be improved. That results especially from the different principles for 5-axis milling (see Fig. 10.7/42/)

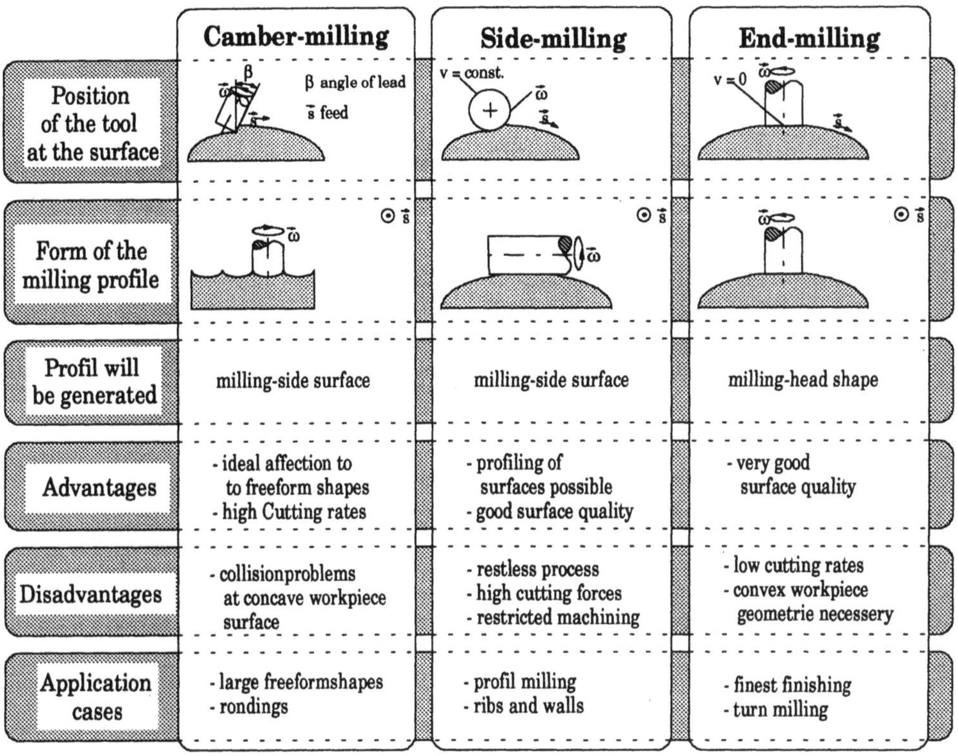

Fig. 10.7. Principles of 5-axis milling

For "camber-milling", the milling tool is directed normally to the workpiece surface and then sloped in the feed direction (lead angle).

By this principle, the centre of the milling tool with the unfavourable cutting conditions (speed 0m/s) is not in contact with the workpiece surface. This technique results an ideal adaptation to freeform shapes, allowing milling with defined and nearly constant cutting and interference conditions. Camber-milling is used for roundings or machining of large freeform shapes.

Side-milling leads to good adaptation of the milling profile to the form of the milling tool. Normally using a constant speed, the surface quality can be very high. Only in the case of necessary usage of long and slender (small) milling tools may the result be a dynamically unstable process (chattering rattling). In this case the milling forces have to be reduced. The main application areas are the machining of high small (and unstable) walls and ribs.

End-milling is used for the machining of convex shapes by the farehead of the endmill. This procedure allows the finest finishing, which leads normally to the best surface qualities.

Based on these general possible principles, the advantages and disadvantages of 5-axis milling procedures are summarized in Fig. 10.8 /42/

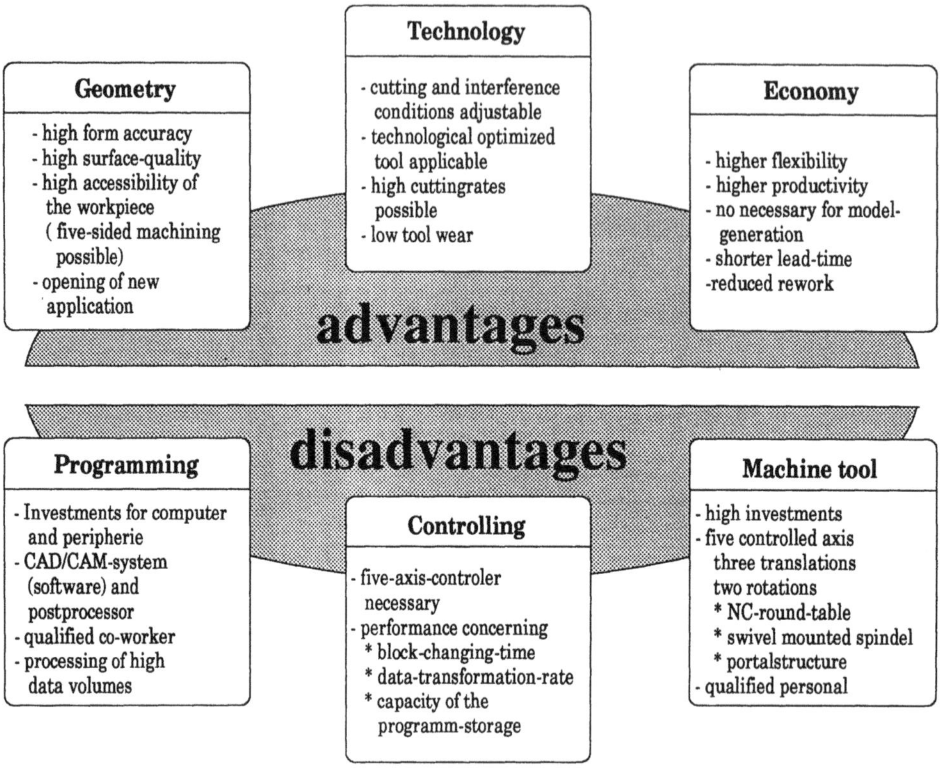

Fig. 10.8. Advantages and disadvantages of 5-axis-machining

The most important advantage relates to the possible complicated workpiece geometry.

By using 5-axis milling, high form accuracy and high surface quality are possible. The expenditure for additional manual work (in comparison with three-axis milling) will be reduced. With the extra degrees of freedom, a nearly universal access to any shape element up to five-side machining is possible.

The important disadvantages are the much higher expenditures for

- programming
- controlling and
- machine-tool construction

So in general, an exact evaluation for the total economic justification is necessary.

For general orientation purpose it can be pointed out that the typical application ranges for 5-axis milling are the following:

- large dies for car bodies
- ship propellers and turbine plates
- wheeles for compressors

In such cases, 5-axis milling is technologically desirable./44/ Restrictions on the application of 5-axis milling are given in all such cases if collisions with walls and ribs close to the freeform shapes have to be avoided. Another danger is given by too high a tilt angle of the milling tool axis.

The best application ranges for 5-axis milling are therefore provided by lightly bent surface shapes, in convex or concave direction (e.g. stamps for deep-drawings for carbody outside parts).

In general, 5-axis milling seems most suitable for larger parts (larger than the available working spaces of SFM equipments). Summing up the characteristics of related process chains in this Chapter, it can be stated that, for a comparison with SFM procedures, high-speed milling with 3-axis controlled machine-tools has to be carefully considered.

11. Decision Making and Economic Aspects

Making decisions about investment in SFM systems or their application can be a big problem, because it is very difficult to calculate the exact advantages (or disadvantages) of such aspects for example, as:

- time to market
- quality improvements

as most important factors for competitiveness.

The decision-making is therefore primarily the responsibility of entrepreneurs or company managers from a strategic point of view. The following explanations will contribute to systematic decision -making from a technological and economic point of view. In this context it is most important to observe that SFM technologies will be considered an essential new element in an entire highly automated factory environment. For efficient utilization, the following general advice can be given:

the application of SFM technology can be most succesful if it is possible to organize close relationships to other essential CAD/CAM-working fields

11.1 Deciding whether to invest or farm out

11.1.1 Characteristics of the problem

When considering given production tasks, the management and technical staff of any company have to face a lot of different requirements. Therefore what is needed is a definition of manufacturing basics among the many-sided technical and informational aids.

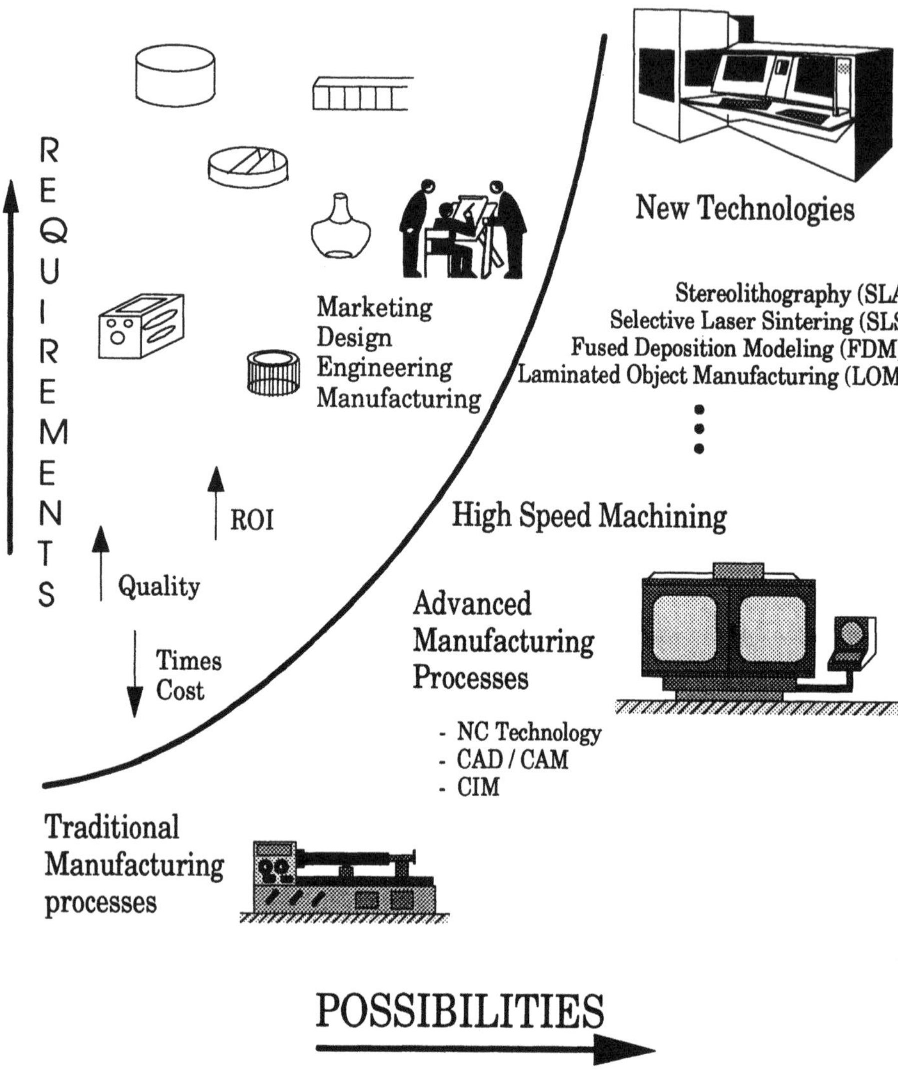

Fig. 11.1. Relationship between Requirment and manufacturing possibilities

For efficient problem solving, more and more new technologies are available. The general problem situation is characterized in Fig. 11.1. These technologies feature different material behaviour, possible accuracies and different possible surface qualities. On the other side are the specific requirements related to the main application areas. Furthermore, the costs, machining times and technological capabilities of the available equipments are also different. So the general question arises as to which is the most suitable equipment for the given company-specific requirements, as well as to what are the suitable application areas in comparison with other manufacturing processes. In any case, the general basic rule for SFM procedures has to be considered too:

- **First, systematize;**
- **Second, automatate!**

11.1.2 Systematic approach for decision-making

The primary goal for evaluating available SFM procedures is to determine the processes best suited for company-specific situations. These are arrived at by considering the following:

- workpiece assortments, including requirements for material, number of workpieces, etc. This has to be determined by a suitable classification system (see Section 11.2)
- availability and access to necessary CAD systems, with 3-D modules and specific interfaces.

Based on the requirements and prerequisites, a systematic approach to decision-making can be carried out in the following steps:

1. Determination of representative parts by consideration of part groups (see Section 11.2). It is advisable to select a part represented in the region of the upper level of its part group (at least larger than 4 inches). This well founded requirement is influenced by the shrinkage factors (see Fig. 11.2).

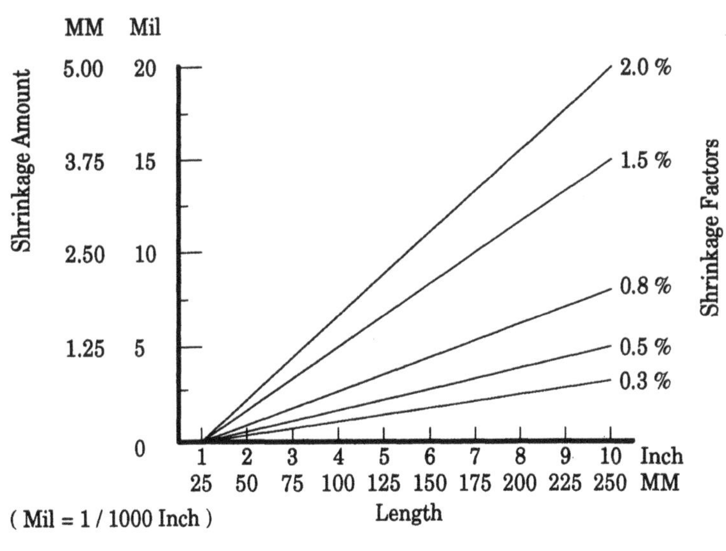

Fig. 11.2. Relation Between part sizes (length) and shrinkage factors

2. Preparation of 3-D CAD data on a suitable workstation (considering the availability of the necessary interface for the anticipated SFM system(s)).

3. Carrying out of a prototype part building procedure with a vendor or service bureau. In this connection it is necessary to determine exactly all time and cost elements for the

    - checking of input data
    - preparation of part building
    - part building procedure itself
    - postprocessing (post-curing if necessary, cleaning, finishing, etc.)

    Therefore, the application of exact process documentation is necessary (see Chapter 7).

4. Checking of the test part(s) with, particular regard to

    - tolerances
    - surface quality
    - form stability
    - curling, etc.

    in agreement with the given requirements.

    An example for evaluation of a selected test part is given in Section 11.2.

5. Based on the selected test part(s), rough calculation (estimation) of the entire necessary time expenditure and capacity utilization.

6. Through consideration of the estimation and calculation of the possible time and cost savings (by comparison with given and currently used procedures; and comparison with other suitable SFM procedures...) decide between

    - in-house investment, or
    - using a suitable service bureau.

11.2 Part classification for SFM

Throughout manufacturing industry there is a basic subdivision into part classes (rotational, box type, sheet metal, etc.). The whole structure of workshops and work division between factories is based on this principle. Independent of the nearly universal application ranges of most SFM procedures, a suitable specific subdivision also seems necessary.

The concrete technological process realization is based on several factors; on the basis of personal experience and confirmed by other research results (i.e.,/57/), the general part structure (overall part appearance) is the most influential factor (see Fig. 11.3).

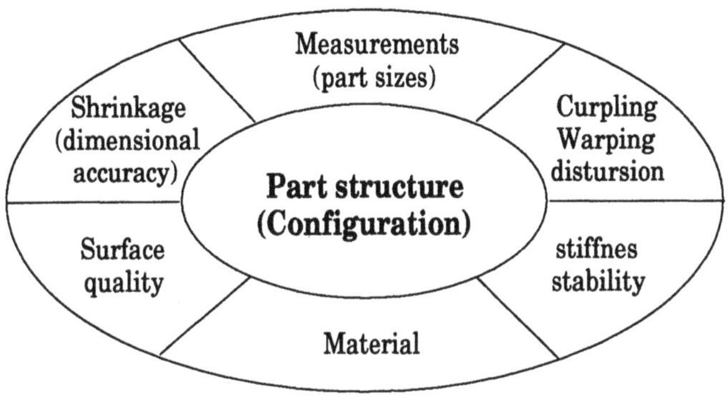

Fig. 11.3. Object related influence factors for quality justified SFM-procedures

Further strongly influential factors are especially

- part size in relation to the dimensional accuracy
- material properties, including strength and stiffness

The available classification methods and schemes for mechanical part manufacturing are not suitable for SFM procedures, because the typical application ranges for these principles are restricted to a higher degree of complexity. Most simple part classes are not suitable for generation by SFM procedures. Therefore a suitable specific part classification is proposed as follows, using the philosophy of Group Technology in a modified form.

The main aspects for a suitable subdivision and classification scheme for SFM procedures are:

- similarities for specific rules for building procedures - dependent on the determined principle
- consideration of typical application ranges for different SFM processes

- general classification independent of other essential influencing factors like materials, measurements (sizes), accuracy, etc.

The result, a classification scheme based on general part structures, is demonstrated in Fig. 11.4.

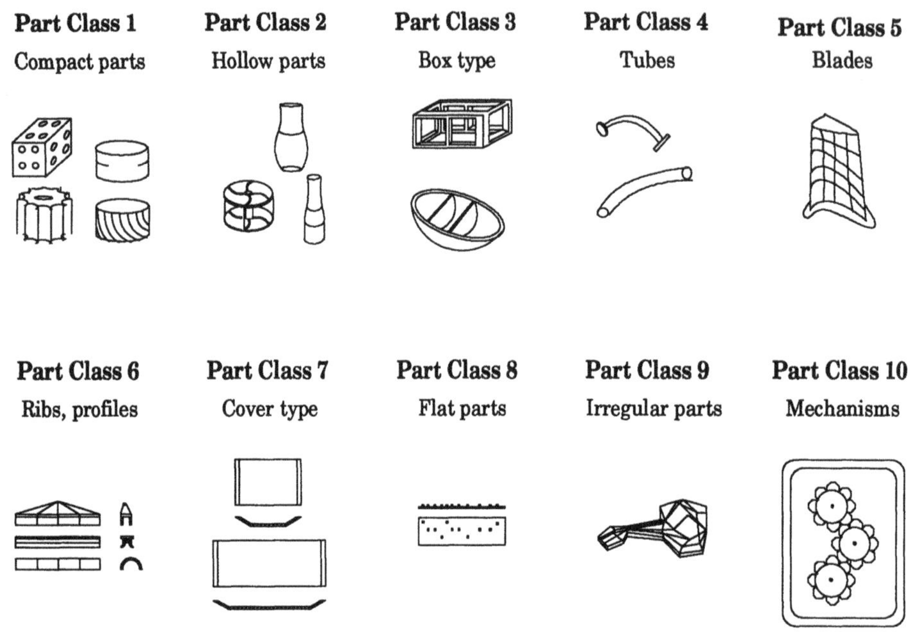

Fig. 11.4. Typical part-structures for SFM-techniques

In this context a further detailed subdivision for every part class is possible depending on

- materials (corresponding to different applications in agreement with Fig. 8.1, see Chapter 8)
- measurements (part sizes)

The part sizes and part structures are most related to shrinkage, distortions and curling effects. Therefore specific experiences have to be acquired and included in the necessary system of rules.

It should be emphasized that this subdivision makes it possible to create a suitable database and likewise also a knowledge base. Furthermore, it becomes possible to make an entire workpiece assortment more transparent for part analysis and selection of test parts. Based on this background, a representative sample part for evaluation of SFM systems can be designed and tested.

11.3 Methodology for execution of SFM Benchmark tests.

11.3.1 General remark

The execution of benchmark test is a traditional practice, and necessary for all kinds of highly productive and expensive equipment (e.g. CAD/CAM workstation, NC machining systems, etc.) Wherever a relatively broad spread of possibilities is offered for specific user requirements, the execution of benchmark tests is absolutely necessary. The dynamic development and increased range of commercially available SFM systems on offer (currently 12 different types of equipments worldwide, partly in different types of equipments, partly in different sizes), mean objective decision-making is essential. There are no comprehensive and general experiences aailable related to technological and economic evaluations. Specific requirements of SFM processes and important aspects for the preparation and execution of benchmark tests will be explained below.

11.3.2 Principal possibilities for SFM benchmark tests

Before a benchmark test for SFM processes is prepared, the available opportunities for advanced manufacturing processes at the factory concerned have to be considered. The new procedure has to be integrated into the entire process chains, and other advanced manufacturing processes must be considered.

Concerning SFM procedures, other competitive processes are 3-,4- and 5-axis milling and especially high-speed machining. After checking out these possibilities, the necessary benchmark test for evaluating different SFM procedures based on the proposed classification system (see above) can be limited to the following types.

1. Benchmark tests with selected special parts (e.g. turbine blades, hollow parts). This is possible if the enterprise has to produce only such selected products, if necessary in different sizes.

2. Benchmark tests with selected representatives of particular part groups. This typical case requires careful part classification and, based on this, the determination of suitable representative parts.

3. Definition of determined complex parts aiming to test the performance of available systems in specific aspects.

The USA experts call such methods "challenge test parts".

The available references include such test parts with details of the "accuracy" explained. Considering different aspects, a specific test part was designed and approved with different partners.

11.3.3 Aspects for determination of a challenge (or sample), test part

11.3.3.1 Main aspects

Currently the restrictions concerning accuracy and curling are the main problems of the available SFM systems. Therefore these aspects have to be considered. Furthermore some other items are essential:

- the test part has to be relatively simple and must be designed with low expenditure.
- the amount of material necessary has to be small
- essential requirements of the entire SFM process chain, especially postprocessing, also have to be considered.

With reference to the general aspects, the test part called "TUD-Phantasie" (see Fig 11.5) was designed. The basic ideas concerning the features used, can be explained as follows.

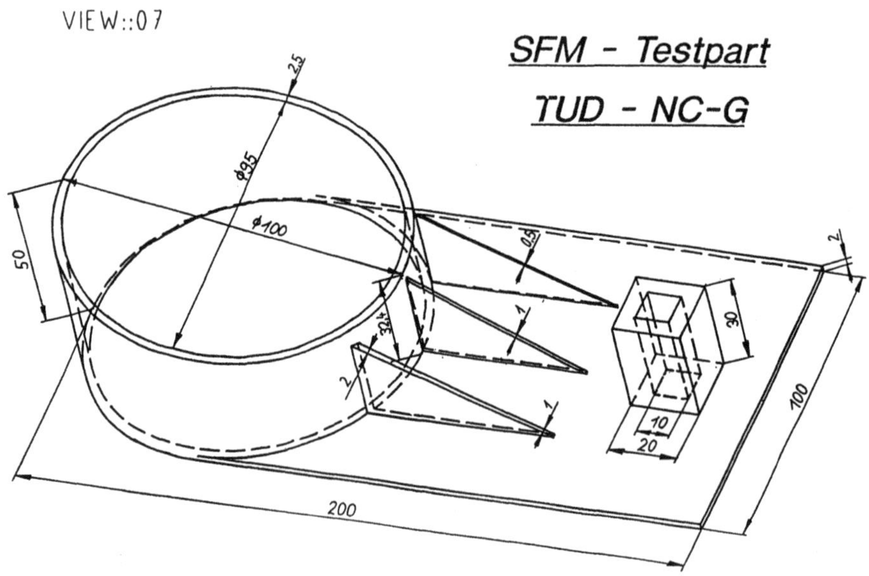

Fig. 11.5. SFM-Testpart TUD-NC-G - Measurements

11.3.3.2 Specific aspects of the TUD. NCG Test part

Based on personal experience, specific problems concerning accuracy and deformation arise especially in the ranges of part classes 6,7,8 (compare Fig. 11.4). Therefore the challenge test part was composed of the following basic elements.

- a levelled, relatively thin-walled foundation plate with measurements 100 x 200 mm (shrinkage and curling happens especially in such sizes)
- a cylinder with closed volume, because the levelling process of the stereolithograpy procedures leads to specific difficulties and additional amounts of time
- small quader also with closed volume, with same requirements and additional effort for cleaning
- small ribs for testing the smallest practically realizable measurements (min = 0.5mm)

11.3.3 Results of the tests

The tests were carried out with well experienced users of all European vendor companies of SFM equipments. It must be emphasized that no single test part is suitable for a general overall evaluation of SFM systems. From consideration of some specific factors it can be confirmed that all European vendor companies can produce objects of the part groups 1-5 (compare fig. 11.4) in a reliable way.

Furthermore the overall complexity of the entire problem has to be considered, which did not allow a complete evaluation.

The essential relevant factors

- equipment and process realization
- experts with specific knowledge and experience

have nearly the same range. So it could be that in a particular case available specific know-how at one equipment balances a material-related - or process-related influencing factor. In other words, the human factor also plays an important role. So the following results are not only an evaluation of technical processes. Based on these remarks, the comparison of the test results in different companies will be done in three ways:

- general evaluation
- geometrical parameters
- economic comparison

General evaluation

The general evaluation was based on the visual impressions: it was noted that each partner was able to realize the test part in good condition. But a nearly 100% ideal solution cannot be confirmed.

Geometrical parameters

The evaluation of the geometrical parameters was based on a specific measurement scheme (see Fig. 11.6). Considering the total realizable accuracy ranges, a measurement accuracy of 1/100 mm and the performance of 1/10 mm is sufficient.

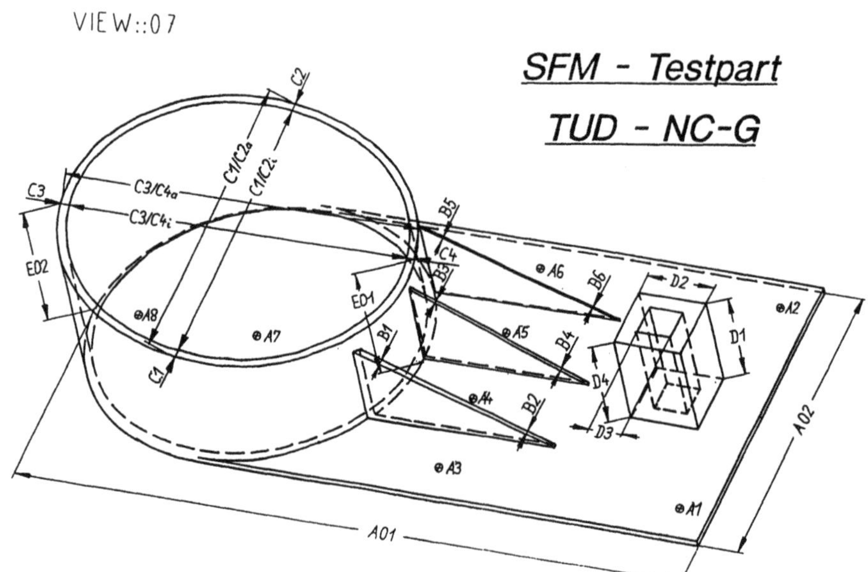

Fig. 11.6. SFM-Testpart TUD-NC-G - Selected measurements

Related to the form features included, the results are subdivided into

- main measurements and
- detail measurements (see Fig. 11.7 a,b).

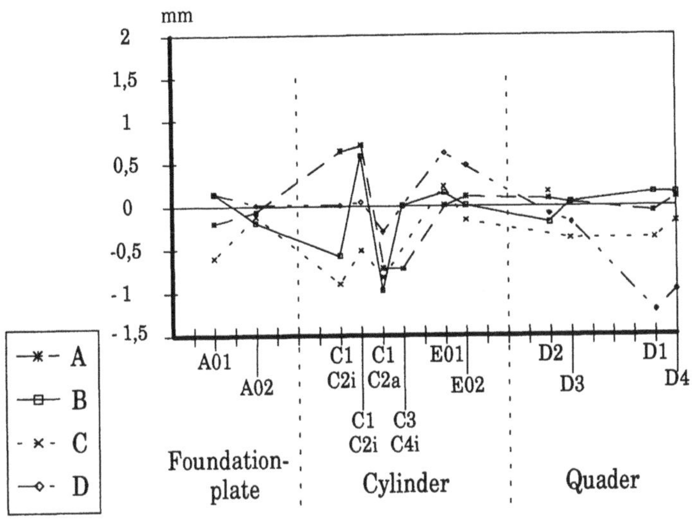

Fig. 11.7a. Main measurements

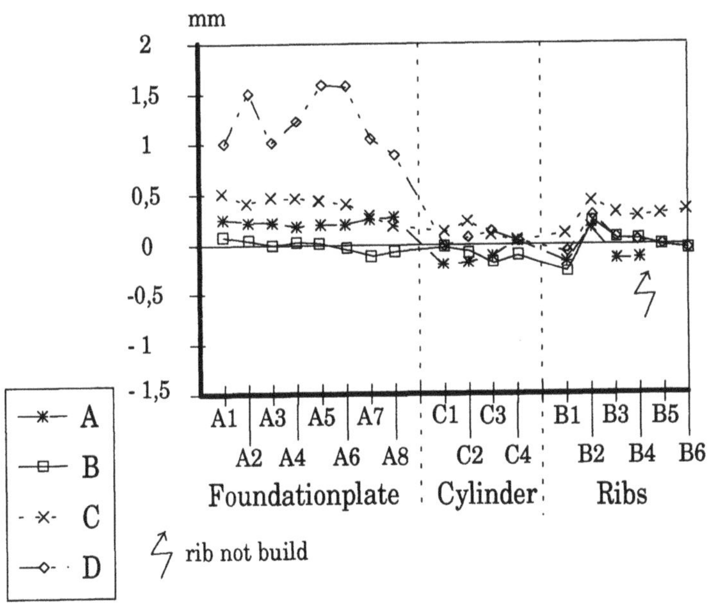

Fig. 11.7b. Detail measurements

The results allow a detailed evaluation concerning the deviations in

- x and y directions and
- the z-direction.

In general, the deviations of all systems included are in an acceptable range. An exception is given by system D, with drastic deviations of the detailed-measurements of the foundation plate. For a final evaluation, the inclusion of economic parameters is necessary.

Because not all detailed economic parameters are available, a first approach is provided by the available building time.

The necessary times for the same sample (challenge) part are shown in Table 11.1.

	A	B	C	D
Data preparation	1 h		3 h	2,9 h
Building time	11 h	7,5 h	8,5 h	25,5 h
Postprocessing	2 h	1,3 h	2 h	0,6 h

Table 11.1. Comparison of essential expenditures

But for an exact cost calculation, and also a final evaluation, all other cost factors also have to considered.

Therefore the other important factors will be explained, in the following pages.

11.4 Economic aspects and criteria for economic justification

The final aim of any kind of evaluation is the exact determination of all essential influencing factors and then, based on the values of each, the resulting cost analysis.

The essential cost element are included in Table 11.2, the necessary input data for which can be obtained from the described process documentation. Additional input data are the cost factors of the equipment employed, as well as information on postprocessing materials and auxiliary materials.

The essential influencing factors, and the methodical steps, will be explained as follows.

Task	**time /h/**	**direct cost**	**material cost**	**other cost**	**remarks**
CAD 3D-modelling	t_{mod} ...	$C_{mod}=t_{mod}(C_{equ}+C_{eng})$ =...(...+...)	-		external workstation
datatransfer model data-checking	t_{check}	$C_{check}=t_{check}(C_{equ}+C_{eng})$ =...(...+...)	-		Computer + (SFM-equipment)
data-preparation (slicing, built param.)	t_{prep}	$C_{prep}=t_{prep}(C_{equ}+C_{eng})$ =...(...+...)	-		Computer + (SFM-equipment)
building procedure	t_{build}	$C_{build}=t_{build}(C_{equ}+C_{eng})$ =...(...+...)	C_{mat}=part / weight*C_{mat}		SFM-equipment
postwork (postcuring, cleaning)	t_{pw}	$C_{pw}=t_{pw}(C_{equ}+C_{eng})$ =...(...+...)	C_{mat}= cleaning mat.		
		$C_{d.c.}$=	C_{mat}=	C_{waste}=	possible range 10-70%
				C_{moint}=	cost per year

Table 11.2. Time and cost factors - SFM-process

11.4.1 Cost Engineering for SFM procedures

Times and costs are vital economic. In some cases, comparisons between SFM systems have been based only on the time for building procedures. It must be emphasized that such a comparison is not sufficient. A better approach and a realistic evaluation is possible from consideration of the following basic principles for cost engineering for SFM.

With the term SFM, will be assumed that all essential economic factors are included. The essential factors are related to the entire process chain and necessary indirect cost-factors /45, 46, 47/.

On this methodical basis we consider such important factors as like user support

- for data handling
- for processing complex STL files or other object interfaces
- for correcting erroneous object files
- for the build data preparation and
- for supervising and controlling the building procedure

by suitable software modules. Additional cost elements for postprocessing work and unplanned costs (waste, maintenance) can be included. Important time and cost factors for different SFM procedures are summarized in Table 11.2. The following explanation for practical usage seems necessary!

11.4.1.1 CAD 3-D modelling

The geometrical modelling of 3-D objects at a suitable workstation can normally be done independent of the SFM procedure. But the determination of time and cost could be of interest in two cases

1. If the full cost calculation for a determined part is requested
2. If different CAD/CAM workstations are available, with different performance. So the comparison for efficient user-support is useful.

11.4.1.2 Data transfer and Model data checking

If the requested interface standards (e.g. STL Files) are available in the correct form, data transfer is possible without problems. But if additional transformations are necessary, for example from VDA-FS or IGES files into STL format, some additional time is necessary. Most important are the utilities for discovering incorrect object models and software aids for correcting small holes in a surface or volume model. In these cases the time t_{check} could be of a significant size.

The cost factors are given as follows

$C_{equ} = [^{cost\text{-}Unit}/h]$ Workstation or other powerful equipment

$C_{eng} = [^{cost\text{-}Unit}/h]$ For this task a highly qualified engineer is required

11.4.1.3 Data preparation

The model data preparation includes in most cases the

- slicing and
- merge procedure, and the
- determination of build parameters

For these tasks, also powerful workstation or computers and qualified engineers are normally necessary.

$C_{equ} = [^{cost\text{-}Unit}/h]$

$C_{eng} = [^{cost\text{-}Unit}/h]$

11.4.1.4 Building procedure

Some aspects to be taken into consideration in relation to the building procedure are:

- the equipment costs may be different, depending on the utilization time (1, 2 or 3 shifts). On the other hand, the utilization time depends on the possibilities of unattended manufacturing during night shifts. The basic cost of equipment has to be included, plus additional costs for communication (e.g. ethernet cables, transceivers...), air-conditioning etc.

- a similar situation applies for the engineering cost. If the SFM equipment operates without a supervisor or skilled worker, the cost factor can be zero. On the other hand, if it is necessary to have two or three operators, this cost factor could be an important one.

- in connection with the building procedure the material cost

  $$C_{mat} = \text{part weight(kg)} * \text{price } (^{\text{cost-unit}}/\text{kg})$$

  has to be considered.

- the part weight can be determined for the most SFM procedures direct from the object weight. One exception is given by the Solider 5600 system. As explained in Section 5.2, the material of the entire vat contents has to be considered (partly resin and partly wax). So a special calculation is necessary, depending on the degree of filling for every procedure (the relation of the total weight of all objects in a determined procedure to the needed material depends on the part of greatest height).

11.4.1.5 Postprocessing work

The expenditure for postprocessing work is different for the varying SFM procedures.

- the cost of equipment C_{equ} is dependent on the necessary devices (e.g. post-curing oven, ultrasonic cleaner, protective equipment, air-pressure devices, etc.)

- if the cleaning procedure of complicated parts is done manually, then the labor cost has to be calculated as an engineering task. This is necessary because the operator responsible for cleaning has to know the determined part with all specific details very exactly.

- the material cost is given by the necessary share of cleaning fluids.

11.4.1.6 Other costs

Essential other costs result from waste and maintenance.

- the expenditure for waste can be an important cost factor if, because of shrinkage, distortion or other failures, more than one attempt is necessary. The possible failurerate can lead to additional costs of 10 - 70 % related to the total cost of the first part (and more!)
- the usual maintenance cost is given by the price of the yearly contract

11.4.2 Overall economic justification

The overall economic justification is connected with good utilization of the capacity of the highly productive SFM equipment. Based on the average times for the selected benchmark test objects, a generalization of the following kind is possible.

It is necessary to determine the approximate number of different object types, related to the part classes, part size ranges and other factors. Based on the overall manufacturing time, it is possible to extrapolate the utilization to the available capacity. This allows the determination of the breakeven point.

If the utilization is not sufficient to justify the installation of such expensive equipment, some other additional typical advantages have to be economically weighted. Many advantages can be quantified for estimation of overall cost savings for the different application ranges. An example is the economic justification arising from successful activities of the marketing department (using "show and feel" objects, without full utilization of the installed equipment).

11.5 Comparison of traditional and new procedures

11.5.1 General remarks

As explained in Chapter 9, the most important "traditional" procedures are the different processes using milling machine tools. These include mainly the milling operations, but also boring, etc. if necessary. Cost elements can be calculated, independent of the specific procedure, because the principal process sequences are nearly identical. Essential process steps for NC milling are demonstrated in Fig. 11.8.

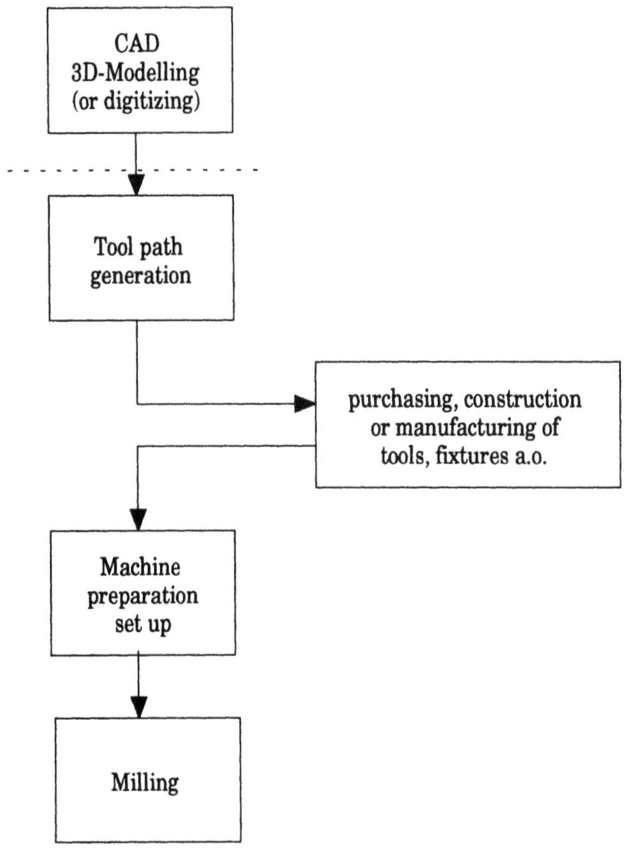

Fig. 11.8. Essential process-steps for NC-milling

The 3-D geometrical modelling is normally the same as for SFM procedures. The possible variant with digitizing is not typical of highly complicated parts. For comparison, the starting point is the expenditure for tool path generation. At most of the available CAD/CAM systems this is a very time consuming task, because it includes several decisions:

- determination of machining strategies
- selection of machining variants
- determination of suitable tools, chucks, fixtures
- determination of suitable working values etc.

An important factor for comparison could be the delay in the process sequence by purchasing or construction, and in some cases the necessary manufacturing, of tools, fixtures, chucks, etc. If special tools and fixtures for determined operations are necessary additional expenditure and in most cases a lot of time could be required. This disadvantage could be of more influence than the pure cost comparison. Also the machine preparation requires some time, especially for exact positioning of the tools used. Most time-consuming are the milling operations themselves, which require much experience and specific knowledge.

11.5.2 Cost calculation scheme

For a systematic approach to cost calculations for the different kinds of milling operations (3- or 5- axis and high-speed milling) the general scheme in Table 11.3 can be used. The specific requirements and differences will be characterized as follows.

Task	**time /h/**	**direct cost**	**material cost**	**other cost**
CAD 3D-modelling	t_{mod} ...	$C_{mod}=t_{mod}(C_{equ}+C_{eng})$ =...(...+...)	-	-
Toolpath generation	$t_{NC\text{-}gen}$	$C_{NC\text{-}gen}=t_{NC\text{-}gen}(C_{equ}+C_{eng})$ =...(...+...)	-	-
purchasing, construction of tools, fixtures ...	$t_{construction}$	$C_{constr}=t_{constr}(C_{equ}+C_{eng})$ =...(...+...)	-	C_{tool} or / and $C_{fixtures}$ or / and C_{chucks}
machine preparation	t_{prep}	$C_{prep}=t_{prep}(C_{equ}+C_{eng})$ =...(...+...)	-	
Milling	t_{mill}	$C_{mill}=t_{mill}(C_{equ}+C_{eng})$ =...(...+...)	-	
maintenance (service)				$C_{indirect}$ (DM/year)

Table 11.3. Essential time and cost factors for milling

11.5.2.1 CAD modelling

There is no difference in the 3-D geometrical modelling, in comparison with the SFM procedures. But there is an important change if the 3-D data are generated by digitizing procedures, in which case a specific comparison is necessary.

11.5.2.2 Tool path generation

Tool path generation requires in every case a powerful CAD/CAM system with a module for tool path determination on a workstation. The necessary cost therefore consists of both cost elements:

- equipment and
- engineering cost

In addition to the general software module for tool path determination, a specific postprocessor is necessary for the determined machine tool. In the case of 5-axis milling machine tools, this is a relatively expensive software component.

11.5.2.3 Purchasing or construction of tools, fixtures etc.

The probability of necessary additional tools, fixtures, chucks, etc. increases with the complexity of the required geometrical objects. That means, in typical cases of highly complicated SFM parts, a comparison with possible milling operations has normally to consider these additional cost elements and also the additional time.

11.5.2.4 Machine preparation

The machine preparation requires a highly qualified skilled worker. In most cases the programmer, that means the engineer, has to give some support in the machine preparation. This can be considered in the cost elements by a factor:

$$\Sigma C_{op3ax} = C_{op} + 40\%\ t_{prep} * C_{eng} \quad \text{for 3-axis milling}$$

$$\Sigma C_{op5ax} = C_{op} + 70\%\ t_{prep} * C_{eng} \quad \text{for 5-axis milling}$$

11.5.2.5 Milling

In the case of 3- and 5-axis machining, it could be, that long machining time allows multi-machine operation. In these specific cases, the cost elements for the operator could be reduced by a determined factor (approximately 60 - 70 %).

11.5.2.6 Maintenance costs

Because of the high prices of all equipments, the maintenance costs have to be considered on an annual basis. Then a realistic cost comparison is possible.

11.6 Organisational aspects

Regarding the organisational aspects, the essential kinds of utilisation and integration into the industrial respectively factory environment are as follows:

- Universal usage for different partners in service-offices

  This seems one typical kind of utilization such high productive new technology. Especially for small and medium sized companies it seems the only possible kind of usage.

  In this case the service-office is requested to fulfil many-sided requirements as it was explained in chapter 8 (application ranges).

  In large enterprises a typical kind of application range is related (connected) to the CAD/CAM-department.

  One of the used terms "direct CAD-manufacturing" is one feature of this development.

  In this case main requirements result from design-verification. The advantage of this solution is given by the excellent model-data-handling by the high qualified CAD-experts. But it is not sure if they oversee all other application-possibilities. Therefore it seems important, to promote by the management the principle of Concurrent Engineering.

- For medium sized enterprises, the best suited and mainly used application area seems the pattern mould and die manufacturing department.

One important aspect for this kind of factory organisation lies in the capabilities of the well experienced engineers and skilled workers. Based on these experiences, they are able in most cases for the best suitable economical solution for the choice between NC-manufacturing or SFM-technology.Furthermore, this kind of organisation allows an efficient work division between the CAD/CAM-department and the pattern, mould and die manufacturing department.

- If the requirement of a medium sized enterprise is relatively manysided and if a good utilisation of a determined SFM-system is in the own company not to guarantee, so the cooperation with different partners (service-office with different equipments) seems the right way.

Because of the different main-requirements, a general orientation for a determined company cannot be proposed.

In every case for the specific decision making the explained technological and economical aspects have to be considered.

12. Further Development Trends

The entire field of SFM technologies is marked by dynamic development in two particular directions:

- continuous improvements of the currently commercial available systems (in this regard note that, most probably, not all applied systems will survive)
- the development and practical application of a remarkable number of new principles.

This application-oriented book therefore includes, mainly, the commercially available systems which allow multiple industrial usage. It should be noted, that the current state of the art is an excellent starting point for further higher developments.

12.1. Improvement of commercially available systems

The drive for continuous improvements of usable processes is directed toward all essential process steps, including

- the performance and user-friendliness of available geometrical modellers and CAD/CAM systems
- simplification of the data transfer by reduction of the necessary amount of data
- software solutions for better understanding of the topological problems of complicated part structures, especially with many part surfaces
- user-support for the computer-aided preparation of the building procedure (e.g. an algorithmn for the automatated correction of holes in the object-surface)
- user-support for the interactive process realization, and development of adaptive controllers
- improvements of the currently usable materials and extensions of the applicable materials (especially concerning ceramics, metal powders and alloys)
- improvements concerning necessary postprocessing tasks (cleaning, hardening, finishing)
- realization of efficient process chains by combination with other manufacturing processes.

These activities include general aspects concerning

- improvements of accuracies and surface quality
- shortening of process times, etc.

12.2. Important further developments

In addition to the commercial SFM systems currently available, there are approximately a further 20 procedures patented, published and under development.

Fig. 12.1 includes some processes that could probably be industrialized during the coming years.

Name	Description	Developer	Developing stage 1992
High temperature SLS	Application of "Selectiv Laser Sintering" for different alloys and metals	University Austin/Texas and DTM/USA	testequipment
3D-Printing	Layer by layer gluing of powder materials with the inkjetprinciple	MIT-Boston / Cambridge USA	testequipment Soligen Inc.
Hot Plot	Layer by layer gluing of sliced laminates for small parts (desc-top)	Sparx AB (Sweden)	Commer-cialization
Dynamat	application of the classical printing methodology by Gutenberg-using metalic materials	SINTEF Trondheim Norway	testequipment
3D-shape-welding	In three robot axis controled distribution of metalic materials	Babcock & Wilcox USA, University Nottingham/UK	? ESPRIT-project
Spray-Metal Deposition	Modification of the metal-spray methodology by material incress Manufacturing	Carnegy Mellon University	testequipment

Fig. 12.1. Some selected procedures for SFM-technologies

A complete survey of future developments has to consider also the further developments of other important manufacturing technologies. Milling operations especially will be of high importance, related to their accuracy and wide choice of materials.

The increased application of high-speed milling extends the traditional borders of the manufacturing technologies introduced.

The most important advantages of the new SFM technologies are the shortening of lead times and the possibilities for improved integration with all other CAD/CAM technologies.

A summarized outlook on the future developments was given by A.L. Cohen /48/.

"Any speculation about the future should begin with a set of assumptions and then attempt to extrapolate their logical consequences. We will take the following assumptions as our starting point in predicting the impact of these new technologies:

- Eventually, processes similar to today's automated, toolless, pattern-less rapid prototyping technologies will directly produce functional parts in production quantities.

- Parts produced in this way will have an accuracy and surface finish equal to those made by machining. Parts will be produced in final or near-final shape.

- The material and mechanical properties of these parts will be identical, or very close, to those of the metals and plastics in common use today.

- The time to produce any part - once the design data are avai- lable - will be a matter of hours."

The impact on the designer and engineer will be an unprecedented freedom, including the following possibilities

- Can increase part complexity without significant effects on lead time and cost

- Can optimize part design to meet customer requirements, without manufacturing restrictions

- Can reduce parts count, by making in one piece parts that are now made from several because of poor tool accessibility or the need to minimize machining and waste. With fewer parts, we can reduce the time spent on tolerance analysis, selecting fasteners, detailing screw holes and assembly drawings.

- Can design parts without regard to draft angles, parting lines or other such constraints.

- Can design parts which cannot easily be set up for machining.
- Can design parts with accurate, large thin walls.
- Can design parts without using stock shapes to minimize machining and waste; can design more organic, sculptured shapes for functional or aesthetic reasons.
- Can design without regard to the particular manufacturing equipment available.
- Can test prototypes which are identical to production parts.
- Can minimize material and optimize strength/weight ratios without regard to the cost of machining.
- Can minimize time-consuming discussions and evaluations of manufacturing possibilities.

Impact on the manufacturer: enormous cost savings

- Can eliminate design, manufacturing, and verification of tooling.
- Can realize profit earlier on new products, since fixed costs are lower.
- Can reduce parts count and, therefore, assembly, purchasing and inventory expenses.
- Can reduce labour content of manufacturing, since part-specific setting up and programming are eliminated, machining/casting labour is reduced, and inspection and assembly are reduced.
- Can reduce material waste
- Can reduce waste disposal costs
- Can reduce material transportation costs
- Can reduce inventory cost for raw stock and finished parts (make only as many as required), therefore:
- Can reduce storage requirements
- Less inventory is scrapped because of design changes or disappointing sales
- Can simplify purchasing since:

unit price is almost independent of quantity, therefore only as many as are needed short-term need be ordered

Quotations vary little among supplies, since fabriction is automatic and standardized

- Can purchase one general purpose machine rather than many special purpose machines, and therefore:

  . Can reduce capital equipment and maintenance expenses

  . Need fewer specialized operators and training

  . Can have smaller production facility

  . Can schedule production with much less effort

- Can reduce inspection reject rate since:

  . Can reduce number of tight tolerances required when parts must mate

  . Can avoid design misinterpretation (instead, "what you design is what you get")

  . Can quickly change design dimensions to deal with tighter tolerances

  . Can achieve higher part repeatability, since toll wear is eliminated

- Can reduce spare parts inventories (produce spares on demand, even for obsolete products).

Impact on the marketer: new capabilities and opportunities

- Can greatly reduce time-to-market, resulting in:

  . Reduced risk: no need to project customer needs and market dynamics several years into the future

  . Products which fit customer needs much more closely

  . Products offering the price/performance of the latest technology

- Can test-market new products economically

- Can change production capacity according to market demand, virtually in real time

- Can increase the diversity of product offerings and pursue market niches currently too small to justify due to tooling cost (including custom and semi-custom production)

- Can more easily expand distribution and quickly enter foreign markets.

Impact on the consumer: choice and low cost

- Can buy products which more closely meet individual needs and wants

  . Can choose from a much wider diversity of offerings

  . Can buy (and even contribute to the design of)

  affordable products built-to-order

- Can buy products at lower prices, since the manufacturers saving will ultimately be passed on.

So in general important progress toward the future can be realized by application of the new technologies.

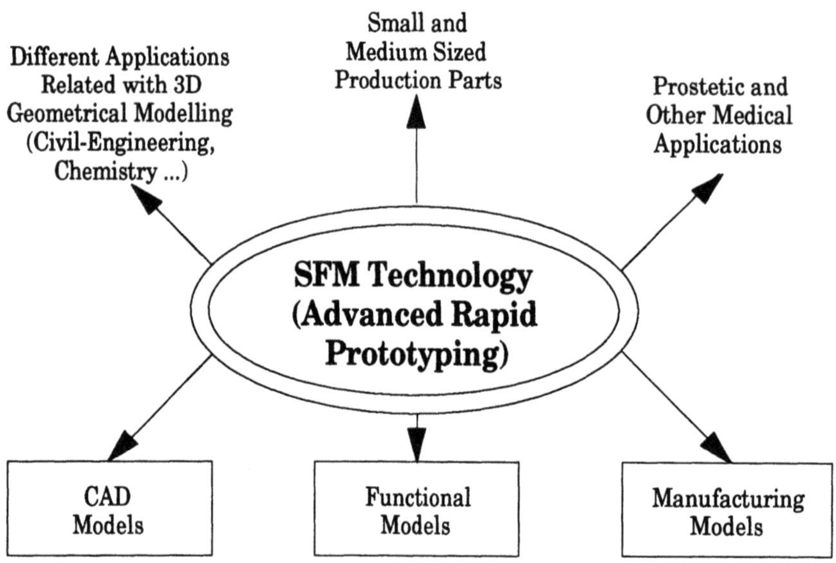

Fig. 12.2. Current and future application ranges of SFM-technologies

References

/1/ Merchant, M.E., "Technology Assesment of the Computer-Integrated Automatic Factory", Annals of the CIRP, Vol. 24, no 2, 573-574, 1975

/2/ Björke, O., How to make Stereolithography into a Practical Tool for Tool Production, Annals of the CIRP, Vol. 40/1, 1991

/3/ Kochan, D., "Developments in Computer Integrated Manufacturing", IFIP-State of the art - CAM, Springer-Verlag, Heidelberg, Berlin, New York, Tokyo, 1986

/4/ Kochan, D., Complex Machining and AI-methods, Proceedings IFIP-Working-Conference, Gaußig, Nov. 1991, Elsevier, North-Holland, 1991

/5/ Kruth, J.P., Material Incress Manufacturing by Rapid Prototyping Techniques, Annals of the CIRP, Vol. 40/2, 1991

/6/ Poindexter, J.W., Rapid Prototyping in an Integrated Product Development Environment, in Second International Conference on Rapid Prototyping, Dayton/Ohio/USA, June 24-25, 1991

/7/ Hull, C., "Apparates for production of three dimensional- objects by Stereolithography", US-Patent 4.575.330, August 1985

/8/ Kodama, "Display 3 dimensional information to a physical formed model", Trans. of Electronics and Communications Society, Vol. 17, no. 6, April 1981 (pp 237-241)

/9/ Andre, J.C. a.o. "Procede et Dispositif pour Realisier un Modele de Piece Industrielle", French Patent 8509055, June 14, 1985

/10/ SLA 250 - User Refernce Manual, 3D-systems, Inc. Valencia Cal./USA

/11/ Lumbye a.o., INSTANTCAM, Study and development of different techniques machine concepts and materials BRITEEURAM, Project No. BE-3527-89

/12/ Dolenc, A., Mäkelä, I., Hovtun, R., Better Software for Rapid Prototyping with INSTANTCAM, PROLAMAT'92, Tokyo

/13/ Jacobs, P.F., Rapid Prototyping & Manufacturing Fundamentals of Stereolithography, Dearborn, SME 1992/USA

/14/ Kirschmann, C.F. a.o., Stereolithography support structure design for Rapid Prototyping in 2. International Conference Proceedings on Rapid Prototyping, Dayton/Ohio, June 24-26, 1991

/15/ Dötsch, E., Stereolithography: Lasergestütztes Verfahren zur Herstellung von dreidimensionalen Modellen, dima 9/90

/16/ Eßer, K.J., Pressegespräch 3D Systems GmbH, Darmstadt, 10.10.1991

/17/ Oberhofer, J.,Stereos - A european approachto Rapid Prototyping, Proceedings Third International Conference on Rapid Prototyping, Dayton, 1992

/18/ Arita, K., Harimata, H., Kuzusako, J.,

Development and Application for high precision Solid Creation System in "Dayton" '92

/19/ Levi, H., "Accurate Rapid Prototyping by the Solid Ground Curing Technology", Proceedings Solid Freeform Fabrication Symposium, Austin, Texas, 1991

/20/ Fudim, E., Rapid Prototyping, Desktop Manufacturing Systems, Light Sculpting Inc Milwonkee, WI USA, 1990

/21/ Kimble, L.L., The Materials Advantage of the Selective Laser Sintering Process, Proceedings Third Solid Freeform Fabrication Symposium August 3-5, 1992 Austin/Texas

/22/ Nutt, K., "Selective Laser Sintering as a Rapid Prototyping and Manufacturing Technique" in Proceedings Solid Freeform Fabrication Symposium 1991, August 12-14 Austin/Texas

/23/ Crump, S., "The extrusion process of Fused Deposition Modeling" in Proceedings Third International Conference on Rapid Prototyping, Dayton, OH June 7-10, 1992

/24/ Feygin, M., Hsieh, B., "Laminated Object Manufacturing: A Simpler Process" in Proceedings Solid Fabrication Symposium, Aug. 12-14, 1991, Austin, Texas, USA

/25/ Sachs, E., Cima, M., Cornie, J., Brancazio, D., and Curodeau, A., Tree Dimensional Printing: Ceramic Tooling- and Parts Directly From a CAD Model, in Massachusetts Institute of Technology, Dayton 90

/26/ Fan, T., Lauder, A., Sachs, E., Cima, M., The Surface finish in three-dimensional printing, in Massachusetts Institute of Technology, Dayton 90

/27/ Masters, W.E., The ballistic particle manufacturing process, 1. International Conference on Rapid Protyping, Dayton , June 1990

/28/ Mueller, T., "Applications of Stereolithography in injection molding", Conference Proceedings in Secound International Conference on Rapid Prototyping, Dayton/Ohio, June 1991

/29/ Schmidt, L.D., Philips, W.L. "Automative Case Study" in Conference Proceedings Succeful Applications of Rapid Prototyping Technologies, April 7.-8. 1992, Dearborn/Detroit, MI/USA

/30/ Clegg, A.J., Precision Casting Process, Pergamon Press, Oxford, New York, Frankfurt, 1991

/31/ Metal Spray Toolmaking Gains, Investment Casting Foothold Thanks to Laser Patternmaking, TAFA Incorporated 146, Pembroke Road Concord, NH 03301-5706 USA

/33/ Lumbye, K.J., Hovtun,R., Making Electrodes for EDM with rapid Prototyping, in Proceedings "Third International Conference on Rapid Prototyping, Dayton, OH June 7-10, 1992

/33/ Pfeiffer, R.W., Stereolithography Accuracy: Influence of Photosensitive Materials, in Dayton 1992

/34/ Gargiulo, E.P., Theory of Stereolithographical Printing, DuPont Company, June 10, 1992

/35/ Thyu, G.E., Computer numerical control of machine tools, Oxford, Newnes 1991

/36/ Kochan, D., Developments in Computer Integrated Manufacturing - State of the art CAM, Springer Verlag Berlin, Heidelberg, New York, Tokyo, 1986

/37/ Weule, H., Konstruktion und Fertigung von Freiformflächen, Tagungsband Karlsruher Kolloquium, 27.-28. Februar 1991, WBK Universität Karlsruhe

/38/ Zeid, J., CAD/CAM - Theory and Practice, Mc Graw-Hill-Inc., New York, 1991

/39/ Bedworth, D.D., Henderson, M.R., Wolfe, P.M., Computer Integrated Design and Manufacturing, Mc Graw-Hill International Editions, New York, 1991

/40/ Klink, Prospects of HSC 1200, Walter Klink GmbH, Max-Eyth-Str. 18, W-7012 Fellbach 5 (Oeffingen), Germany

/41/ Brüggemann, U., Konstruktive Anforderungen an Fünfachs-fräsmaschinen, Werkstatt und Betrieb 124 (1991)6 pp. 473-476, Carl Hauser Verlag München

/42/ Feldermann, J. Fünfachsiges NC-Fräsen, VDI-Z 133 (19916 pp. 75-87

/43/ Spur, G., Potthast, A., Wojcik, L., Erweiterungs-möglichkeiten einer CNC-Steuerung für die 5-Achsen-Fräsbearbeitung, ZWF 86 (1991)6 pp 273-277, Carl Hauser Verlag München

/44/ Cubital proposels RP benchmark files Rapid Prototyping Monitor Volume 1, No. 4, January, February 1992

/45/ Humphreys, K.K., Katell, S., Basic Cost Engineering Marcel Dekker, Inc. New York ans Basel

/46/ Guthmann, N., How to keep Product Costs in line marcel Dekker, Inc. New York and Basel

/47/ Malstrom, E.M., Manufacturing Cost Estimating Marcel Dekker, Inc. New York and Basel, 1981

/48/ Adam, L. Cohen, Founder & Editor Rapid Prototyping Report, Los Angeles, Proceedings Third International Conference on Rapid Prototyping, Dayton,OH June 7-10, 1992

/49/ Lawry, M.L., I-DEAS Student Guide Integrated Design Engineering Analysis Software; Structural Dynamics Research Corporaton 2000 Eastman-Dr. Milford, OH 45150

/50/ Joshi, S.R., Miller, R.A., Kinzel, G., Rapid Prototype Development by Part Decomposition Method; Net Shape Manufacturing Report No ERC/NSM-D-92-19 April, 1992; The Ohio State University, College of Engineering

/51/ Lowrey, R.L., Rules for Solid Modeling AUTOFACT'90, Nov. 12.-15., 1990, Detroit, Michigan, SME

/52/ Sowar, R.M., Geometric Modeling ... A FOundation for Application Integration, AUTOFACT'90, Nov. 12-15., 1990, Detroit, Michigan, SME

/53/ Wozny, M.J., "Data Driven Solid Freeform Fabrication" in Proceedings PROLAMAT'92, Tokyo, June 1992

/54/ Rock, S.J., Wozny,M.J."A Flexible File Format for Solid Freeform Fabrication", Solid Freeform Fabrication Symposium 1991, Marcus, Beaman, Barlow, bourell, and Crewford, eds., Published by the University of Texas at Austin, August 1991

/55/ Rock, S.J., Wozny, M.J., Generating Topological Information from a "Bucket of Facets" Solid Freeform Fabrication Symposium 1992, Marcus, Beaman a.o., University of Texas at Austin, Ausgust 1992

/56/ User's manual for DuPont SOMOS^{TM} 3110 Solid Free-aging Photopolymer (Version 1.2). The DuPont SOMOS^{TM} Material Group, New Castle Corporate Commons, 1992

/57/ Wall, M.B., "Making Sense of Prototyping Technologies for Product Design ". Master of Science Thesis, MIT 1991

/58/ Kochan, D., Innovative Produktentwicklung und Fertigung mittels Solid Freeform Manufacturing, 1. Hearing NC-G S. Okt. 8, 1992 Ulm/Germany (in german)

www.ingramcontent.com/pod-product-compliance
Ingram Content Group UK Ltd.
Pitfield, Milton Keynes, MK11 3LW, UK
UKHW050945290926
15470UKWH00005B/1646

9 780444 896520